T0389167

DIALECTICS OF KNOWING IN EDUCATION

Dialectics of Knowing in Education strengthens the philosophical basis of formal education that has been weakened by neoliberalism over the past 30 years. It theorises and encourages human existence based on social action, culture, inquiry and creativity so that citizens in democratic association can formulate their own understandings of the world and be their own philosophers of practice.

Under neoliberal capitalism, formal education has become a key economic driver and factor for all countries, but has exacerbated social division and inequality. This has led to an increased pressure on education systems to emphasise individual gain and prosperity at the expense of community care and concern. Drawing on the work of Dewey, Mead, Freire and Biesta, the author argues that formal education at all levels must be transformed so that it does not seek to impose knowledge and truth, but situates knowledge as being constructed by democratic learning circles of staff, students and citizens.

Focusing particularly on the notion of praxis and specific issues involving Indigenous, feminist and practitioner knowing, this book will help scholars, practitioners and policy makers to transform their education theories and practices in ways that encourage democracy, emancipation, social action, culture, inquiry and creativity.

Dr Neil Hooley is an Honorary Fellow in the College of Arts and Education, Victoria University Melbourne, Australia. He has interests in Democracy and Social Justice, Philosophy of Education, Pragmatism, Critical Theory and Action Research. He supports recognition, respect and reconciliation between the Indigenous and non-Indigenous peoples of Australia.

DIALECTICS OF KNOWING IN EDUCATION

Transforming Conventional Practice into its Opposite

Neil Hooley

LONDON AND NEW YORK

First published 2019
by Routledge
2 Park Square, Milton Park, Abingdon, Oxon OX14 4RN

and by Routledge
711 Third Avenue, New York, NY 10017

Routledge is an imprint of the Taylor & Francis Group, an informa business

British Library Cataloguing in Publication Data
A catalogue record for this book is available from the British Library

Library of Congress Cataloging in Publication Data
Names: Hooley, Neil, author.
Title: Dialectics of knowing in education : transforming conventional
practice into its opposite / Neil Hooley.
Description: Abingdon, Oxon ; New York, NY : Routledge, 2018. |
Includes bibliographical references and index.
Identifiers: LCCN 2018015190| ISBN 9781138311930 (hbk) |
ISBN 9781138311947 (pbk) | ISBN 9780429458552 (ebk)
Subjects: LCSH: Education—Philosophy. | Democracy and education. |
Education—Social aspects.
Classification: LCC LB14.7 .H67 2018 | DDC 370.1–dc23
LC record available at https://lccn.loc.gov/2018015190

ISBN: 978-1-138-31193-0 (hbk)
ISBN: 978-1-138-31194-7 (pbk)
ISBN: 978-0-429-45855-2 (ebk)

Typeset in Bembo and Stone Sans
by Florence Production Ltd, Stoodleigh, Devon, UK

CONTENTS

FIGURES

TABLES

ABOUT THE AUTHOR

Dr Neil Hooley is an Honorary Fellow in the College of Arts and Education, Victoria University Melbourne, Australia. He taught mathematics and science at the secondary level for many years before lecturing in higher education. He has interests in the philosophy of education, American Pragmatism, critical theory and pedagogy, participatory action research and inquiry learning as they apply across all areas of knowledge and the curriculum in schools and universities. He has been involved in projects that investigate professional practice, community partnership and praxis learning for preservice teacher education to pursue social justice and educational equity for all students. In addition, he has participated in projects concerning narrative inquiry as research methodology and curriculum construct in primary and secondary schools. Dr Hooley is committed to reconciliation between the Indigenous and non-Indigenous peoples of Australia and sees progressive educational reform as a step towards this end. He strongly supports partnerships between schools, communities and universities as democratic means of improving dignified social life and of learning from and theorising social and educational practice to challenge organisational structures and personal understandings.

PREFACE

Dialectics of Knowing in Education is a manifesto for practitioner-theorists who are restless and impatient for a better world. It strengthens the radical and progressive philosophical basis of formal education that has been seriously weakened and narrowed by aggressive neoliberal ideology around the world over the past 30 years. It argues that a new defining narrative of knowledge and schooling is required similar to those of freedom, peace and equality, for example those that animated the international peace movement during the Cold War period, the civil rights movement in the United States and opposition to the Vietnam War. Such narratives explain and demonstrate what it means to be human. Accordingly, the book draws upon Greek and European philosophy that asked, 'How should we live?' and European Enlightenment that considered, 'What can we know?' to question today, 'What does it mean to experience mind, to act, think, know and create ethically?' This is the question of human subjectivity, of humanity itself, of what it means to act, think and be. Consequently, understanding the human mind, or at least having a more nuanced and consensus model of how it engages and makes sense of its surroundings is essential for a coherent and comprehensive approach to formal education at all levels. In the three decades following World War II, public education was located within a broad social contract that emphasised productive citizenry and personal growth for all. Today, with neoliberal capitalism rampant across major and minor economies alike, the notion of creative knowledge construction for everyone regardless of socio-economic background and based on family and community interest, has been replaced by imposed and conservative appreciations that support private and individual gain. Under these circumstances, schooling is inherently frustrating and alienating for vast numbers of children as they are systematically removed from the big ideas and practices of history and knowledge of which they and their communities are a part and are inducted into a technical and superficial rationality of human existence.

Part I of *Dialectics of Knowing in Education* begins with a review and commentary of materialism and dialectics as originally raised by Greek philosophy and continued by Hegel, Marx and Engels. This has been a consistent thread in philosophical thought over many centuries as human purpose and existence has been interpreted in contrasting idealist, materialist and scientific terms. Discussion then turns to the Aristotelian notion of *phronesis and praxis* as a guideline for living well and for the central idea of education and learning for all citizens in the modern world. Praxis is proposed as a framework of inquiry, ethical conduct and practice, not to be taught as such, but to be experienced as social life and education proceeds. With major and progressive philosophical ideas outlined, Part II focuses on the question of knowledge and how we currently understand the human mind connects with the social and physical environments and makes sense of what is considered to be real. This generates many philosophical and metaphysical problems regarding epistemology, thinking and thought, human consciousness, emotion, desire and the nature of language. Linguists such as Chomsky, Pinker and Wittgenstein offer insights as the question of mind and therefore how we learn and comprehend remains unresolved. Part III discusses some specific issues of philosophy of mind involving Indigenous, feminist and practitioner knowing. While each offers a distinctive perspective, it is argued that there are significant commonalities that suggest universal grammars or modes of understanding across social groups although mediated by economic, political and cultural impulses.

Through this broad ranging, non-linear, reflexive narrative, discussion and inquiry, *Dialectics of Knowing in Education* falls within the broad philosophical tradition of American Pragmatism including over the past century and more theorists such as Dewey, Mead, Vygotsky and Freire. In many respects, the book is a reintroduction of Dewey and Mead in particular to current generations of educators at all levels. It envisages humanity as evolving from continuous cycles of *practice-theorising* to the creative, ethical and intellectual life known today. It defends the democratic process of acting, thinking, inquiring and creating as an absolute human right that should pervade all forms of social and personal being and organisation. Written as a personal narrative, the book describes a genuine process of my inquiry around historical, epistemological and social questions that form the context of formal schooling and education. European and North American thinking and philosophy constitute the major influence of notions of social class, democracy and equity that must frame systems of education and schooling, although other philosophies such as Indigenous and feminist are integrated. The narrative account and description of my own confrontation with these questions will be framed by literature, cases, electronic posts and exemplars of knowledge culminating in a flexible framework for similar reader and practitioner consideration. In this regard, each chapter ends with a case of experience connecting with and illustrating chapter content and challenging readers to relate to their own experience. As well, at the end of each part, a draft or indicative knowledge exemplar will be offered describing thinking and insights to that point of the narrative. It is intended that the knowledge exemplars will act as a 'reflexive interlude' for readers, rethinking

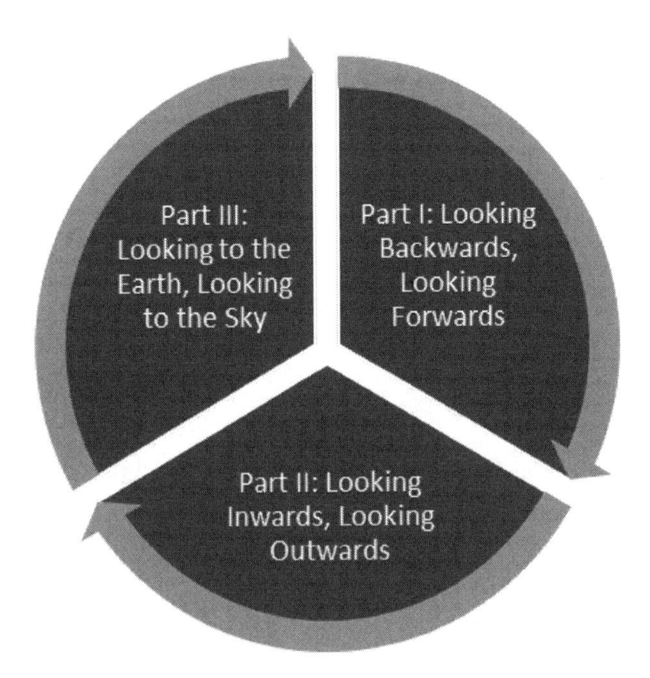

FIGURE 0.1 Structural organisation of narrative, discussion and inquiry

their personal connections with the narrative overall. Figure 0.1 shows this integrated intention:

As mentioned and to illustrate my thinking at particular points of the narrative, a series of cases and schema have been included throughout. A case is defined as a descriptive account of a situation that embodies one or two key principles of social practice without attempting to provide answers or lead the reader to a specific conclusion. Each case is based on a personal experience of mine, but some contextual detail has been elaborated with artistic concession to emphasise the main points concerned. For the purposes of this book, a schema is defined as the organisation of informal thoughts and ideas that are relevant to the narrative at that point in the chapter. Each scheme as part of an overall schema of practice is authentic thought that I have communicated to colleagues via social media at the date shown. Cases and schema are an aspect of narrative methodology that encourage readers to participate in considering and critiquing the issues being raised in a cyclical manner as they construct their own understanding of the discourse to that stage and provide a continuing 'point of return' as the narrative unfolds. In that sense, a narrative is not linear, moving from issue to issue in step-wise form, but enables events of past, present and future to combine and interact with each other.

Dialectics of Knowing in Education extends previous writing and exploration (Hooley, 2010, 2015, 2018) of radical and progressive educational possibilities under neoliberal economies. It draws upon a small number of themes and principles that have emerged from practice-theorising of teacher education in Australia and internationally. These themes and principles include social class understandings of philosophy, society and knowledge; democratic social and educational practice, phronesis and praxis; composition and relation of lifeworlds; centrality of consciousness and language for human existence and learning; human subjectivity and intersubjectivity and, as noted above, the practice of practice-theorising for all citizens. We can draw upon the American Pragmatists here in thinking about such themes as falling within the broad notion of 'society' and how humans act to construct meaning. It is argued that neoliberalism has little interest in such matters, or interprets some as being present and supported by current structures and resources for the benefit of those who exercise power within neoliberal capitalism. As an example of how this process works, the book discusses how international and national testing regimes act in the mutual interests of the minority not only in the specific nature of test items, but in how results are scaled and reported. Figure 0.2 indicates this relationship.

As noted above, *Dialectics of Knowing in Education* seeks to provide a philosophical and reflexive practice narrative that extends beyond historical questions such as 'How should we live?' and 'What can we know?' to pose today, 'What does it mean to experience mind, to act, think, know and create ethically?' These are essential questions for systems of formal education, especially schooling where children are concerned and, in many respects, have not been resolved satisfactorily. They are also essential questions of knowledge, or epistemology, as to how societies, ideologies and schools envisage knowledge and its formation. Under neoliberalism and given its ideological connotations, education has become a site of intense political and economic contestation, with the purpose and organisation of education constantly argued. In the wealthier countries with expectations that most students will complete secondary schooling, billions of dollars are expended each year and the strong supporters of neoliberalism anticipating appropriate return on investment. 'How should we live?' is an ethical question ranging from an individual and self-interested perspective, to a collective and public orientation. 'What can we know?' raises issues regarding the nature of the universe, the place

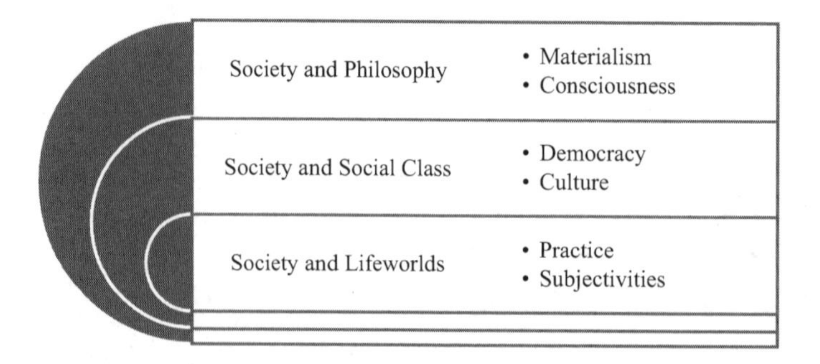

FIGURE 0.2 Themes and principles underpinning human existence

of humans within the cosmos and how we come to understand through human practice, reason and consciousness. 'What does it mean to experience mind?' challenges the notion of meaning that arises from experience and ultimately, how we come to an appreciation of what we call 'mind' as the essence of ourselves. Clearly, the concept of 'self' is important here, where 'to act, think, know and create' are suggested as the basis of self and of awareness. Each term needs to be examined in detail. Finally, humans have evolved the notion of an 'ethical' life, a way of evaluating what we do as being appropriate or not and in the interests of others rather than being purely indulgent. There are various approaches to ethical conduct that will be discussed later. It is not possible for this book to answer these questions definitively, but as we attempt to unravel some philosophical considerations and struggles about them, there will hopefully be deeper engagement with the problems and contradictions involved. This is the process of philosophical practice that is patently absent from mainstream education around the world today. Neoliberal capitalism has no interest in considering purpose and direction other than economic that serves the elite of social power and in strengthening that power regardless of consequences. War, aggression and racism results. It is not being argued here that transformed systems of schooling that focus on personal and community knowledges, language and cultures as the basis of understanding the world will bring about peace and justice in our time. That is the idealist view of history. However it may assist all students as citizens in their process of becoming more human, more compassionate and more ethical as the basis of their active struggles with history, power and exploitation.

ACKNOWLEDGEMENTS

I acknowledge the Elders, families and forebears of the Indigenous peoples of Australia. I recognise that the land on which we live, meet and learn is the place of age-old ceremonies of celebration, initiation and renewal and that the Indigenous people's living culture has a unique role in the history and life of Australia. We live and learn together in the interests of peace and justice for all peoples.

More than can ever be known, I greatly value the support and friendship shown by close friends and colleagues throughout difficult and very challenging neoliberal times when company, care and concern are most appreciated and on other occasions when professional and personal life is slightly more comfortable. Over more recent times, I am therefore most grateful for the memorable discussions, mutual experience, cherished friendship and constant encouragement of friends and colleagues Marie Brennan, Tony Edwards, Sunny Gavran, David Jones, Claire Kelly, Tony Kruger, Catherine Manathunga, Oksana Razoumova, Kerry Renwick, Mary Weaven, Jo Williams and Lew Zipin. Such collegiality has strengthened and enabled the social and educational perspectives and theorising expressed in this book. I also convey heart-felt gratitude for their advice and support over many years to esteemed and principled colleagues Martin Andrew, Julie Arnold, Dorothy Bottrell, Martyn Brogan, Peter Burridge, Efrat Eilam, Gwen Gilmore, Mat Jakobi, Kim Keamy, John McCartin, David Miller-Stinchombe, Brian Mundy, Greg Neal, Peta Oates, Mark Selkrig, Sarah Tartakover and Simon Taylor. I am in personal and intellectual debt to all.

My professional and personal life has been substantially enhanced by the theorising and perspectives of John Dewey, George Herbert Mead, Paulo Freire and Lev Vygotsky and, in more recent time, Gert Biesta. They encourage human understanding through personal and social action and respect the culture and experience that all humans bring to learning about and engaging with the world. Unfortunately, it is this very question that has proven too difficult for many education systems and schools in many countries, especially in the face of highly conservative ideologies that exclude and discriminate. Together, we shall continue the struggle for democratic and critical public education as a basic human right for all citizens.

Neil Hooley
Melbourne
March 2018

DEDICATION

To reintroduce John Dewey and George Herbert Mead to current generations of educators for their historical, social and educational scholarship and efforts in creating a better world through drawing on the best critical and communicative aspects of modern philosophy and modern science:

> Consciousness, an idea, is that phase of a system of meanings which at a given time is undergoing redirection, transitive transformation. The current idealistic conception of consciousness as a power which modifies events is an inverted statement of this fact. To treat consciousness as a power accomplishing the change, is but another instance of the common philosophic fallacy of converting an eventual function into an antecedent force or cause. Consciousness *is* the meaning of events in course of remaking: its 'cause' is only the fact that this is one of the ways that nature goes on. In a proximate sense of causality, namely as place in series history, its causation is the need and demand for filling out what is indeterminate.
>
> (Dewey, 1958, p. 308)

> Characters and the things in which they are embodied endure only in perspectives. This is most strikingly evident in the characters of motion and rest. Whatever is at rest in one consentient set is in motion in another. But this is also true of other characters. There is no sensuous character that is absolutely changeless. We recognise that even mass changes with motion. The general statement for this is found in the resolution of all the physical characters of things into energy, which in every element of a system is constantly varying. Nor can we find in the laws of nature as they appear in experience any absolute endurance. All the enduring relations have been subject to revision.
>
> (Mead, 1938, p. 112)

POEM

Intellect

> Brains envisaged as filing cabinets
> alphabetical order of neat envelopes
> retrieved and opened upon request
> for confronting mysterious realities
> linear events defining conclusions,
> many others not fitting the mould
> embodied networks of understanding
> forming patterns of constant change
> interacting with thoughtful meanings,
> messy connections rather than precision
> enable children to laugh at their easels.

Note. All poems in this book, including *Intellect* above, at the head of chapters and at the conclusion of Chapter 12, are the work of the author, Neil Hooley. They are intended to connect with one of the main themes of the book and of each chapter and to illustrate the integrated knowledge of creative human thinking. Like all art, the poems exist in the eye of the beholder and are designed to challenge interpretation and meaning.

PART I

Looking backwards, looking forwards

Whether or not the objects of existence are real, are real but are not as they appear, or are not real is a little difficult to prove. However it seems that educators around the world are not too concerned with this problem and generally in the neoliberal era proceed without questioning the assumptions of knowledge, curriculum and pedagogy. *Dialectics of Knowing in Education* proposes a systematic way of considering human existence that draws from American Pragmatism and the principles of understanding that arise from social practice. Beginning with the notion of 'looking backwards, looking forwards' enables detailed reflexive contemplation of key philosophical ideas and actions regarding the nature of the universe and of humanity and subsequent problems and issues that arise and must be confronted for democratic improvement. For example, if American Pragmatism suggests that human knowing occurs through living in and with our social worlds, then it follows that staff and children should be enabled to follow such practices in formal schooling as well. Part I of *Dialectics of Knowing in Education* begins therefore with a review and commentary of materialism and dialectics as originally raised by Greek philosophy and continued by Hegel, Marx and Engels. Such work influenced the development of modern science as well as the theories and thinking of Dewey, Mead and others. Materialist dialectics and the attendant notion of contradiction has been a consistent thread in philosophical thought over many centuries as human purpose and existence has been interpreted in contrasting idealist, materialist and scientific terms. Discussion then turns to the Aristotelian notion of *phronesis and praxis* as a guideline for living well and for the central idea of education and learning for all citizens in the modern world. Praxis is proposed as a framework of inquiry, ethical conduct and practice, not to be taught as such, but to be experienced as social life and education proceeds. It can be seen as a continuous act of practice-theorising. Humanities and sciences are discussed as major areas of designated knowledge, their similarities and differences and how they can be considered from a pragmatist standpoint. With major and progressive philosophical ideas outlined, educators are in a much better position to debate and theorise their own practice and to construct equitable learning and teaching environments that are congruent with the way in which learners of all ages and all backgrounds go about investigating meaning.

1

MATERIALISM AND DIALECTICS

> As the morning stillness is transformed
> multiple blooms wait expectantly
> dependent on the mix of activity within
> relations made, unmade and adjusted
> causations of old and new reoriented
> creating new lifeworlds of meaning.

From my early days as a secondary school teacher of mathematics and science, I came to the somewhat personal view without extensive discussion with others, that the distinction between theory and practice was false. This would have been encouraged I suppose by thoughts of what was called 'pure and applied knowledge' and how these should be taught in school. After a lifetime, I still cannot explain exactly why I adopted this position, but I suspect it had something to do with growing up in a country town, close to a temperate green farming environment and an ocean beach where I spent most of my time, winter and summer alike. It is difficult to not wonder about the formation of trees, rivers and waves when they constantly and involuntarily fill the mind and imagination. My formal studies of science and chemistry and work as an industrial chemist probably strengthened these views when I found myself thinking systematically and deeply about the nature of the universe and the interactions of matter. We all make these types of decisions, those that are difficult to defend and, in this case, I can't really remember having any doubt. That is, my view of the universe centres on the combinations of matter and energy and the rearrangements that occur perhaps over very long periods of time. I have never felt the need to invent something else to explain the inexplicable, to verge into the mystical or metaphysical, but I accept that these considerations are important aspects of the great human need to understand. When I came across the writing of Karl Marx (1954/1974, p. 29) at university his words seemed to sum up what I had been observing and thinking over previous years:

> My dialectic method is fundamentally not only different from the Hegelian but is its direct opposite. To Hegel, the life-process of the human brain i.e. the process of thinking, which, under the name of 'the Idea,' he even transforms into an independent subject, is the demiurge (creator) of the real world and the real world is only the external, phenomenal form of 'the Idea.' With me, on the contrary, the ideal is nothing else than the material world reflected by the human mind and translated into forms of thought.

I was excited by this insight, but there were many new challenges here for me to confront, in fact a new direction for my life. I had not come across the notion of 'dialectic' before, nor the German philosopher Hegel, but the concept of the 'material world' being 'reflected' in some way in the human mind to produce 'forms of thought' established the connections between theory and practice. Marx uses the words 'brain' and 'mind' above, an important distinction that I will return to later. I should point out that I was not predisposed to a religious view of the world, perhaps due to the dominance of the natural environment in my daily experience and concepts of the 'creator' were not very significant for me. I was instead fascinated by how combinations of matter and energy hung together and accepted the scientific theory that something like the 'Big Bang' was the starting point. I understood that a problem science has along with all other aspects of social life, is being able to critique its own theories and practices using the same theories and practices that it applies to its own studies. This process of reflexivity can be extremely awkward to implement, especially with questions of ethical conduct that are difficult to define and realise with integrity; it demands understanding and rigour in thinking about how we make decisions, look for alternative explanation and judge what is correct. Whatever the case, Marx suggested to me that I was a materialist and dialectician.

Introductions to Hegel and Marx

My understanding of dialectics from Hegel (1770–1831) was somewhat hazy but provided a way for thinking about thinking that appealed to me. I understood that Hegel was an important figure in German 'idealism' that includes Immanuel Kant, Johann Fichte and their struggle with notions of *a priori* and *a posteriori* knowledge, meaning knowledge that is known independently of experience and knowledge based on experience respectively. Hegel considered that the items of experience could only be known by the way the mind considers them, or that items of experience are unknowable, existing as ideas only. In this manner, thoughts within the brain were not the abstracted pictures of actual objects and processes, but such objects and processes were the realised picture of the idea. Under one definition, dialectics is an approach to knowledge, discussion or argument involving different points of view that ultimately leads to truth, or at least, new understanding. Hegel had a more detailed consideration. He emphasised the 'whole' or 'totality' of existence, indicating that the stages within the whole are only partially true or untrue. This means that the whole subsumes within it all the stages that have occurred previously, described by the German concept of *aufheben*, to 'rise up' or in English, 'to sublate'. For example, a stone when raised from one position to another, has a new position and possesses different kinetic energy, but remains the same stone, the stone does not disappear. This is the process of 'negation' whereby the second state negates the first as a step towards wholeness. In the first relationship of 'being' between seemingly opposed concepts, the concepts appear to have 'nothing' to do with each other. In the second relationship and with further analysis however the concepts imply each other, or without one we do not have the other. Third, we reach a higher form of understanding that may still be unclear or contradictory but offers new possibilities. This process of 'being, nothingness, implication, being' is often referred to as 'thesis, antithesis, synthesis', although this has been criticised as missing the detail outlined by Hegel and the relationship that exists between the three moments. For me, Hegel was providing important guidance about human thinking with concepts of wholeness, contradiction, dialectic, negation and relationship, rather than a more static and linear view of how we come to know and become.

Karl Marx (1818–1883) was influenced by the work of Hegel and was a member of a group or movement known as Young Hegelians. These philosophers and writers rejected the idealism of Hegel and his view that the contradictions of history had ended resulting in the Prussian state. Marx instead looked to materialism including the work of Ludwig Feuerbach. In his 'Thesis on Feuerbach' published after his death, Marx famously proclaimed (Engels, 1976, pp. 61–65):

> The chief defect of all hitherto existing materialism – that of Feuerbach included – is that the thing, reality, sensuousness, is conceived only in the form of the *object or of contemplation*, but not as *sensuous human activity, practice*, not subjectively. Hence, in contradistinction to materialism, the *active* side was developed abstractly by idealism – which, of course, does not know real, sensuous activity as such.
>
> Feuerbach wants sensuous objects, really distinct from the thought objects, but he does not conceive human activity itself as *objective* activity. Hence, in *The Essence of Christianity*, he regards the theoretical attitude as the only genuinely human attitude, while practice is conceived and fixed only in its dirty-judaical manifestation. Hence, he does not grasp the significance of 'revolutionary', of 'practical-critical', activity.

In this statement, Marx unambiguously identifies human beings as evolving through *sensuous human activity, practice*, conceived as *objective reality*. He comments that this is in marked contrast to the idealist conception of history involving the 'theoretical attitude' (of mind, *a priori*) that recognises human practice contrasted in religious terms between Christianity and Judaism, 'fixed only in its dirty-judaical manifestations'. Finally, in relation to human change and progress, he states that the idealist view cannot explain 'revolutionary' thinking with 'practical-critical' connections. Marxism then considers the main features of dialectical materialism as including:

- nature as a connected and integrated whole, in a state of constant movement and change that occurs, not as a simple repetition of what has passed, but as an upward and onwards movement, from a new state to another, from the simple to the complex;
- contradictions as inherent in all phenomena with a positive and negative side such that different aspects of the contradiction are becoming and dying away enabling quantitative features being transformed into qualitative features and vice versa.

From these considerations, it is but a short leap to see how these ideas and principles allow us to study and analyse social life and history of society, known as historical materialism. In general terms therefore, I summarise historical and materialist dialectics as the science of general trends, contradictions and laws of motion and development of nature, society and knowledge. Dialectical and historical materialism demonstrates the philosophical difference between materialism and idealism and therefore engages a radical rather than conservative view of what it means to be human. Central to materialist philosophy is the view that human consciousness (see Chapter 6) is created through active engagement with the social and physical worlds over time and is not preordained. To try and illustrate these ideas a little more, I suggest the guides shown below, together with a brief note at this stage to orient our thinking to schooling, in preparation for future chapters.

1. Unity of opposites

Definition. Internal aspects of a phenomenon, object, or idea that are interrelated and are in constant contradiction and motion.

Example 1. Positive and negative charges within an atom or molecule.

Example 2. Public and private orientations of society.

Example 3. Recognition within schools that all students have different experiences that produce different ideas in transition.

Note. Relationship between theory and practice raises questions regarding unity of opposites. Theory and practice or theorising and practising can be seen as the one entity with the balance between them shifting depending on the conditions. A theory/practice entity rather than dualism changes the way we view each other – and how we view teaching and learning in schools.

2. Qualitative/quantitative change and transformation

Definition. Change that occurs between qualitative aspects (internal characteristics: structure, connectedness, value) and quantitative aspects (external characteristics: size, volume, scale) of a phenomenon, object, or idea.

Example 1. Continuous germination of seeds into flowers, plants (evolution also example of quality to quantity)

Example 2. Conversion of labour into commodity value and human value.

Example 3. Social class assumptions within schools regarding test results and student capability.

Note. Inclusion of the history and philosophy of science into both science and humanities classes enables all students to gain some understanding of dialectics and how science continues to change and develop, perhaps a unity of opposites in itself.

3. Negation of the negation

Definition. Development of the new from the old through current aspects of the unity of opposites being transformed or negated into new aspects.

Example 1. Chemical reactions creating new products not existing before (chemistry also example of transitions from quantity to quality).

Example 2. Class antagonisms creating new social and economic arrangements not existing before.

Example 3. Schools negating the true nature of mathematics and then being negated by socialist understandings of mathematics.

Note. Including the history and philosophy of mathematics into all year levels enables children to experience that mathematical knowledge occurs in the same way as other knowledge and that they can investigate and create mathematical understanding in the same way for themselves. Teachers can allocate half their mathematics sessions each week to project work and philosophical investigations that will in fact enhance the other half of the week devoted to depersonalised, traditional, predetermined truths and calculations.

Connecting with the Greeks

It may have been better if I had started reading about Greek philosophy a number of years before Hegel, Marx and the European Enlightenment, but life usually does not follow an ordered pattern. But when I came across the work of Socrates, Plato and Aristotle, together with many others, I was impressed. In the first case, I admired the questions they were asking and challenging others to consider. Many of these problems remain unanswered today and, although we now accept that mistakes were made, I saw logic in their thinking. Second, and as a teacher of mathematics and science, I respected their attempts at grasping the nature of knowledge and the distinction noted above between senses and reason. This showed me that the divide between theory and practice will not be easy to resolve. In the third case, I was aware that I was becoming involved with primarily European philosophy and that there were other philosophies that I needed to study. However my family background is essentially Irish and with the strong connections between the UK and Australia, I thought that it was appropriate to at least have a good understanding of the Europeans before I diverged into other areas.

From my experience and reading, I supported the view that the dialectical materialist view of knowledge places practice in the primary position and that 'Practice is higher than theoretical knowledge, for it has not only the dignity of universality, but also of immediate actuality' (attributed to Lenin as cited by Mao, 1968, p. 4). This approach proposes that there is an initial observation of the external relations of an event, followed by a sudden cognitive leap to the internal relations that are present, a new understanding that is consolidated by cycles of practice moving from perception to conception. In the broadest sense, given that there is 2,000 years of experience between their thinking and vastly different societies, Aristotle and others laid the basis of the practice approach to knowledge supported by Marx and Engels. For example, Aristotle noted that, 'action and its contrary cannot exist without thought and character' (Barnes and Kenny, 2014, p. 118) and added that:

> Thought, as such, however, moves nothing, but only that which aims at an end and is practical; for this rules the productive intellect as well, since everyone who makes for an end and it is not matters of production which are ends *tout court* (though they may be ends for something or someone) but rather matters of action.

He then goes on to state that, 'Knowledge, then, is a state capable of demonstrating things' (Barnes and Kenny, 2014, p. 119), emphatically bridging the theory practice divide. I found this line of thinking remarkable and exciting, particularly given my involvement in modern science at this stage and my attempts at coming to grips with what the universe was all about. Many other Greek thinkers of around this time also made wonderful contributions to knowledge such as Democritus, who thought about the smallest particles of matter called atoms, or Heraclitus, who like Hegel later, considered 'wholeness' and 'change' as involving the eternal and unifying movement of opposites, the basis of contradiction. I am not exactly sure why we have not reached agreement on these matters over 2,500 years, except perhaps that they ultimately concern how we understand ourselves as tiny specks in the universe, something that even the deepest introspection may not be able to achieve at this time. Humans have of course, when attempting to explain the inexplicable, found recourse in imagining something else, another force or entity in the cosmos. This may involve a soul that connects with gods, or the construction of a narrative of beliefs and rituals that explain origin and existence. In my view, I am prepared to accept the main elements of what we now call scientific explanation of universal beginning and progress, bearing in mind the many gaps and uncertainties that still plague our journey.

Dewey and American Pragmatism

I came across John Dewey (1859–1952) quite by accident. I was browsing the education section of a university bookshop near where I lived and discovered a book entitled *Democracy and Education*. It was a pity that lecturers and colleagues had not alerted me before this to a philosopher of such historic importance, particularly given my interest in the links between theory and practice and what this meant for formal schooling. I well remember coming across the following passage (Dewey, 1916, p. 87) and without exaggeration, it changed my life:

> Since a democratic society repudiates the principle of external authority, it must find a substitute in voluntary disposition and interest; these can be created only by education. But there is a deeper explanation. A democracy is more than a form of government, it is primarily a mode of associated living, of conjoint, communicated experience.

Dewey was explicitly bringing the concepts and practice of democracy and education together. This opened my eyes to the nature of democracy at a more fundamental level than its usual connotations of voting within various organisations, significant and necessary as that may be, but as a function of how we live – humans as democratic beings. He was saying that we are not governed by external authority but are governed by ourselves in the way that we come to learn and understand and make our own decisions. I took this to mean that as an organism we experience the world in a democratic way, whereby we act, think and realise for ourselves, influenced by others but not dominated by them; what I think is my decision to make. That is, Dewey (1958) made a distinction between this process being entirely individual and living, learning and communicating with those around us in a collaborative progression of exploration and knowing. In my later reading, I saw that his definition of 'progressive' education was based on this idea, of humans living with and in the world throughout their lives in continuing cycles of activity, communication and reflection, of progress, rather than in a more 'traditional' sense of accepting what others have proclaimed. A dialectical and materialist view of nature and history that recognised change as constant certainly fell into the progressive category.

John Dewey belonged to a school of philosophy that became known as American Pragmatism. Not surprisingly, American because it originated in the United States from the 1880s and because it put forward an approach in contrast to some major ideas of European or Continental philosophy at the time. Pragmatism from the Greek word 'pragma' meaning action, interestingly made a connection with Greek thinking and emphasised the centrality of human action in creating sense and thought. Charles Sanders Peirce is credited with initially raising the question of pragmatism formally which was then developed by William James, Dewey, George Herbert Mead and Jane Addams. Dewey, Mead and Addams all worked in Chicago and were close colleagues. Addams was the founder and co-ordinator of Hull House, a refuge or settlement residence for refugees and destitute families, particularly mothers and children. Dewey often visited Hull House and I suspect there were similar principles and practices being implemented at the schools he established at the University of Chicago. Jane Addams was active in the anti-war movement of World War I, opposing any American involvement and was a key figure in the suffragette movement; she was awarded the Nobel Prize for Peace in 1931. Peirce published a well-known paper entitled 'How To Make Our Ideas Clear', in which he outlined some of the main principles of pragmatism. He suggested three levels of engagement from which the clarity of ideas emerge. First, a familiarity with the idea or object; second, a need to understand the issue such that it can be used and discussed and third, observation of what the idea or object does in its totality. He wrote (Peirce, 2015, p. 31):

> It appears, then, the rule for attaining the third grade of clearness of apprehension is as follows: Consider what effects, that might conceivably have practical bearings, we conceive the object of our conception to have. Then, our conception of these effects is the whole of our conception of the object.

In this way, Peirce is outlining the very basis of pragmatism. He is not only saying how we make our thoughts and ideas clear for others, but how we clarify our ideas for ourselves. In other words, this is how we forge our human relationship with the world. He went on to describe how this occurs when attempting to understand the concepts of hard, weight and force that are central concerns of physics. For example, we might observe and become aware of force at dinner time in relation to slicing a loaf of bread or pushing a plate across a table. Parents might tell their children to be careful, to not be so vigorous and not to hurt themselves. Children might not have a word for the process, but with experience and discussion, come to realise that they can create change, be useful, but must be careful when bringing this process, called force, into play. Dewey added an essential point regarding knowledge to this concept of pragmatism, when he wrote in Democracy and Education (Dewey, 1916, p. 344):

> The theory of the method of knowing advanced in these pages may be termed pragmatic. Its essential feature is to maintain the continuity of

> knowing with an activity which purposely modifies the environment. Only that which has been organised into our disposition so as to enable us to adapt the environment to our needs and to adapt our aims and desires to the situation in which we live is really knowledge.

Again, we encounter here a principle that identifies pragmatism and knowledge, that of 'continuity of knowing' such that we actively change the environment for our purpose. Note that this is a two-way process, where we act on the environment and the environment acts on us. Based on these theories of pragmatism (and I would argue, dialectical materialism), Dewey and colleagues developed the philosophy of 'inquiry learning' for schooling in the early part of the twentieth century, for which there was much comment and criticism. William Kilpatrick, Teachers College Columbia University, advocated 'project-based' pedagogies as a means of implementing inquiry learning, a process that is still adopted by many schools around the world today. More discussion of pragmatism including its criticisms will occur later in the book, but for now, I want to briefly mention two other major historical figures who have contributed enormously to the epistemology of a sensuous, democratic, dialectical and ethical humanity.

Lev Vygotsky (1896–1934) was a Russian psychologist, working on a new theory and approach to psychology following the 1917 Russian Revolution. This project involved embracing the key ideas of Marx and applying them to how we think and understand, in other words to conceptualise human consciousness in new ways. As mentioned above, this takes up questions such as the wholeness of humanity with the universe and nature, materialism, theory and practice, production and learning. Vygotsky considered that this should be done by understanding the socio-cultural and economic context in which humans lived, rather than concentrating overly on the inner workings of the brain. Dewey's writing tended to focus on the intellect, although comments regarding social and communicative life can be found throughout his scholarship, as indeed by his comments on democracy and associated living. Development of Vygotsky's approach was cut short by his early death, but his insights continue to be researched and actualised to frame progressive educational practice world-wide. He noted the importance of 'historical materialism, which explains the particular significance of the abstract laws of dialectical materialism for a particular group of phenomena' (Moll, 2014, p. 5, as cited by Veresov, 2005, p. 36). Progressive educators often bracket Dewey, Vygotsky and Paulo Freire together as the bulwark against the excesses of neoliberalism and conservatism in education. Freire (1921–1997), in coming from Brazil, offered a Latin and South American perspective not only to education for the people generally, but for literacy in particular. He developed the notion of 'critical pedagogy' in working on literacy in the countryside involving groups of people in 'learning circles' investigating and resolving issues of concern for the local community. This process was 'dialogical' in that discussion needed to be entirely respectful, non-coercive and directed at consensus among all involved. In summing up this approach, Freire may be best known for his concept of 'conscientisation', or the process where 'knowing subjects achieve a deepening awareness both of the socio-cultural reality which shapes their lives and of their capacity to transform that reality' (Freire, 1972a, p. 51). We could interpret this definition as 'critical consciousness'. It is possible then to draw a clear line from the Greeks and their questioning of senses and reason, to Hegel and the concepts of contradiction and dialectic, to Marx and the linking of theory, practice and production and now to Dewey and colleagues, Vygotsky and Freire and the philosophy of American Pragmatism, inquiry and conscientisation. Progressive education therefore has an impressive ancestry extending back centuries, linking the great puzzles of humankind and providing an extensive network of understandings regarding modern schooling.

Discovering Biesta

I mentioned earlier that over the past 30 years or so, neoliberal capitalism as a highly successful ideology, has squeezed out philosophy in favour of sociology (White, 2018). It is not surprising that what I would call sociological movements and identity politics such as feminist, environmental, anti-racist, multiculturalism and sexual orientation have come to the fore as communities and individuals struggle to find their voice and dignity within economic determinations. My suggestions for coping with problems of the role – indeed decline – of philosophy in education today, are outlined in my book *Learning at the Practice Interface* (Hooley, 2015). An example that is relevant here of one trend swamping another is the discussion by Rorty (1979/2009, p. 5) of the history of philosophy, where he comments that 'we should see the work of the three most important philosophers of our century – Wittgenstein, Heidegger and Dewey' as located within the changing historical context. This historical writing and analysis and critique of 'philosophy as epistemology', is said to have reintroduced Dewey and American Pragmatism to the discipline of philosophy after many years of war, depression, cultural change and globalisation. Rorty also suggested that Dewey set about to 'construct a naturalised version of Hegel's vision of history', although all three philosophers came to realise that a 'foundational' view of knowledge that can accurately represent the real, was not possible. As a new teacher and without knowing a great deal about philosophy and its twists and turns, my reading of Dewey made immediate sense and I could see his descriptions all around me, not only in classrooms but in my experience as well. He seemed to work from what I was observing, feeling and trying to work out and was providing ways of thinking about what was happening. I was imbued with the notion of inquiry that fitted so well with my science classes and which challenged me to structure my mathematics teaching in the same way. Then, after many years of coming to grips with philosophy, as seeing myself as a pragmatist of sorts, I came across the work of Gert Biesta.

I first discovered Biesta's book *Beyond Learning* (Biesta, 2006) when I was thinking of a follow-up to my book on Indigenous education (Hooley, 2010).

Following Dewey, I had constructed my thoughts on Australian Indigenous schooling within an inquiry and democratic framework and my electronic search must have picked up on Biesta's subtitle, 'Democratic Education for a Human Future'. I was impressed with a philosopher of education who not only seemed to be working within the tradition of progressive democracy and of Dewey himself, but also was citing thinkers such as Hannah Arendt in trying to understand the 'democratic person' and a compassionate 'human future'. My mind returned to writers that I had met along the way such as the British philosophers Hirst and Peters and their description and discussion of education, including ethics and education. On a conference trip to London, I had the privilege and opportunity of meeting Gert Biesta and of briefly discussing with him some aspects of Greek philosophy including the notion of praxis that we were considering in Australia. Over a period of 10 years or so, Biesta has published a series of books on the philosophy of education (Biesta, 2006, 2010, 2013) that have positioned him as the pre-eminent philosopher of education in the world today. He has lectured, written and published extensively on a wide range of philosophical questions that will be referenced throughout this book. Of one particular significance for myself and the profession, is his scholarship regarding the pragmatist, George Herbert Mead and the importance of subjectivity and intersubjectivity for education. In his current writing, Biesta challenges conventional notions of teaching and learning, indeed, whether learning itself should be the outcome of teaching. I interpret this to mean learning that is specified by the 'learning industry' within narrow confines. Instead, Biesta advocates that education and schooling is a process of becoming, of becoming more subjective and more human, of becoming another person. Biesta draws upon a range of philosophers, theorists and writers to inform his views who, in my estimation, envisage a conception of humanity that is evolving, knowledgeable, dignified and optimistic; but first and foremost, active. These are difficult ideas and ideas that need an explainable programme of implementation if they are to obtain general recognition. What this body of work does however is provide a philosophical and systematic critique of formal education under neoliberal capitalism, identifying weakness and contradiction that cannot meet the needs of ordinary families the world over.

Thinking about the philosophy of education

My journey to the world of philosophy as sketched very briefly above, may seem somewhat confusing, if not of little significance, but it raises many fascinating questions for me and for education. I'm not sure why philosophy has attracted me so strongly, except perhaps it begins to grapple with epistemology and the nature of knowledge that has puzzled thinkers for so long and, of course, the practice of social inquiry seems to be not only what we do, but how we are. It is unfortunate after so many years in education as a student, teacher and academic that no-one to my recollection has ever encouraged me to delve into the mysteries of human endeavour and essence in an orderly manner, to attempt to understand the connections between Enlightenment figures of for example Kant, Fichte, Hegel and Marx; I had to do this for myself. I will take up the issue of consciousness and language later, but I have come to distinguish these myself as key ideas regarding the human species and for which, practising teachers need at least a scaffold for guidance in their classes. We can refer to Hegel here, although his broad description is open to interpretation (Hegel, 1977, p. 111):

> Self-consciousness is faced by another self-consciousness; it has come *out of itself.* This has a two-fold significance: first, it has lost itself, for it finds itself as an *other* being; secondly, in doing so it has superseded the other, for it does not see the other as an essential being, but in the other sees its own self.

Hegel is proposing that consciousness and self-consciousness do not exist independently but arise in relation to other consciousnesses; existence is social and relational rather than private and absolute. If they do exist independently, then where did they come from, how were they formed? Such a relationship makes one consciousness vulnerable or subject to change upon interaction with an *other*, it becomes *lost* and uncertain compared to its initial state. One self-consciousness can interpret or disregard the other and instead, see the other in its own image or disposition. This explains the sensation in conversation with a colleague or friend, when there is a particular moment or 'feeling' of closeness and understanding, or a 'feeling' of frustration and disagreement. Such a moment of meeting between two aware human beings engages the self-consciousness of each and generates the possibility of new thought and understanding. What could be more important for the teacher, than to appreciate that a process like this could occur with every encounter in class and that this is the genesis of learning, indeed ultimately of humanity?

Let's come forward a few centuries from Hegel and consider some of Biesta's thoughts regarding Dewey. In discussing what he called the 'Communicative Turn' in Dewey's writing, Biesta (2006a, p.31) comments:

> In *Democracy and Education*, the theory of communication not only figures in Dewey's account of how meaning can be communicated, it also provides the framework for a social or communicative *theory of meaning* itself. While participation in a joint activity is central to Dewey's account of communication, he emphasised the importance of the role played by things – both the things around which action is co-ordinated and the sounds and gestures that are used in the co-ordination of action.

This passage indicates the place of human action, participation and communication in meaning such that, as Biesta goes on to discuss, we come to understand objects (ideas, artefacts, practices) not through specific characteristics of the objects themselves, but the use to which objects are put and which we can then engage.

We can work with symbols such as the letter 't' and the numeral '3', but these need to be associated with objects and use before meaning can be ascribed, for example the words 'tree' and '3 trees'. Shared understanding comes about through attaching meaning to objects through action enabling sensible communication to occur. I need to repeat here that we are working with broadly descriptive explanation of human cognition in the same way as Hegel's contribution above and it is still not possible to go to a deeper level. Some advances are being made in neuroscience and brain scanning, but these theories are still at an early stage of development. As educators, what aspects of philosophy we each accept is up to us and our individual world views. In this chapter, I have attempted to outline some of the main ideas that make sense to me and to have highlighted continuity within philosophy that has taken me to Dewey in particular and then Biesta. In his recent work, Biesta discusses issues in art education and utilises Joseph Beuys's art performance of *How to explain pictures to a dead hare* that I will ponder in more detail later. Given I have just used the word *sense* in relation to myself above, let me note a comment by Biesta (2017, p. 31) in which he proposes that 'Teaching is therefore fundamentally a triadic act in which there is *someone* showing *something* to *someone* else'. In describing this process as 'an act of turning another human being towards the world', Biesta is defining sense and subsequent understanding as that occurrence when the other person 'comes into dialogue with the world'. If we put aside the question of 'world' for the moment (which world/s, whose world/s), I see a connection with Hegelian dialectic above and his description of the interaction of consciousness. What this means I think is that philosophy continues to explore and explain what it means to be human and to find a language that provides more profound clarity. For me, Dewey in particular and now Biesta have opened my eyes to a pragmatist (and 'pragmaticism', see Peirce, 2015a, p. 255) vision of knowledge, learning and teaching that guides my social practice every day, that enables me to interact with others humbly, honestly and respectfully as I attempt to live in and with what I understand as world and hopefully, provides a basis for ethical action and introspection. As best I can make sense of this tentative construction, I see this as the dialectical materialist understanding of the universe and of humanity.

Case 1. Golf and experience

It was still cloudy and a little misty as Geoff wheeled his bike out the front gate, but he hoped that it would clear to a fine day later. He had not been to a golf tournament before, even though he had often jogged around the course and its picturesque setting. Growing up in a country town by the beach, Geoff knew that the local golf course had a wide reputation and a number of holes were in the dunes with views of the ocean. Although interested, he had never taken up playing the sport mainly because of the cost involved in buying the equipment and in paying fees to join the club. Unfortunately his school did not offer golf as an elective on sports day. He had often wondered however what it must be like to drive a ball down the fairway, get out of the rough and bunkers, chip to the green and putt to the hole. How do they know how to hit the ball the length needed and where they want it to go, he wondered, it must be a great feeling when it works. I'd like to try that, he thought. Geoff noticed the clouds were lifting as he parked his bike just inside the main entrance and followed the crowds to the first tee. He wasn't sure who the players were, but he had heard on the radio that there were some top golfers competing from around the country, as it was a regional championship. It was impossible to tell what was going through their minds as each player teed off, but Geoff saw that they were concentrating hard and were very serious as they approached the ball and looked ahead to the contours and hazards of the course. After the first few holes, Geoff positioned himself at various vantage points so he could watch the different strokes involved and how each player adapted to the conditions, especially when the ball landed near a tree, or in the scrub. If he was going to have a go at this game, he needed all the information he could get. At the eighth hole, Geoff ran to the green and sat down as near to the edge as was allowed. He had a good view as the three players arrived and considered where each ball had landed. One player was side on to Geoff, but he thought he noticed something as the ball approached the hole; he wasn't sure, but it seemed the ball moved off a straight line and just missed dropping in with that 'clunking' sound. He had never seen real-life putting before and it was different to what he had thought. By now the second player was about to putt and Geoff had a great view from nearby and behind, a clear line of sight to the hole. Amazingly, the ball was hit a few metres to the left of the hole and then, equally amazingly, as it slowed, it changed direction, and dropped in, clunk. Geoff was staggered, he had not realised until that very moment that golfing greens were not flat and that players had to judge where to hit and at what speed so that the ball would 'weave' its way to the hole! How did they do that, what were they thinking when they stood over the ball, how did they know how hard to hit the ball and where? Geoff was still excited but somewhat confused by what he had seen as he collected his bike for the ride home. It must be a lot of practice he thought, but how do we practice thinking, especially chipping and putting? He decided there and then he would have to have a go to find out.

2

SCIENCES, HUMANITIES AND PRAXIS

> Feeble light heralds the dawn
> as remnants of the departing storm
> shudder at horizon's boundary line
> replaced by tributaries of blue and purple
> separating immense bundles of grey
> that form and reform independently.

My views on materialism and the presence of an objective, real world must be shaken of course by how modern science now considers or is reconsidering how the universe is constituted. I remain open to debate, but unreconstructed at the moment. Einstein's spectacular interrelationship between matter and energy for example brings our understanding of each into question, let alone the actual process of how one is converted into the other. I accept the dialectic but that doesn't make it easy to understand. These are conceptual issues for humans to grasp, similar problems to quantum mechanics that attempt to describe the universe 'below' matter at a different fundamental level of information, a confusing and weird mix of probabilities, entanglement, waves, strings and counterintuitive relations. Davies (2014, p. 87) states the challenge as follows:

> Given that the universe could be otherwise, in vastly many different ways, what is it that determines the way the universe actually is? Expressed differently, given the apparently limitless number of entities that can exist, who or what gets to decide what *actually* exists? The universe contains various things: stars, planets, atoms, living organisms . . . why do those things exist rather than others?

It seems to me that the universe could be completely different with a few slight changes here and there, let alone a few slight changes to human sensory perceptions.

We can argue whether the human-derived laws of mathematics and physics define the universe, or whether there would be different laws if the universe was different. Would beings elsewhere in the universe construct space craft in an entirely different manner based on the laws that they have formulated? As Eagleton (2016, pp. 12–13) so aptly puts it, 'Human beings are outcrops of the material world, but that is not to say that they are no different to toadstools', their 'materiality' he continues is of another form. These types of questions have shaped the basis of modern science, even how we define and understand science itself. During the Industrial Revolution and European Enlightenment, there was a combination of empirical and religious viewpoints, but as the notion of the natural world took over as an ordered, clockwork mechanism, the need for a purely theological understanding declined. Hence the separation of science from religion at the time. My own views on these affairs emerged from the cultural context of a country town in regional Australia.

I was probably not cogitating on the nature and relationship of matter and energy when I was running along the beach before a swim, or riding my bike out to the river on a sunny afternoon to try and glimpse the elusive platypus, but I still have vivid memories of that experience. Naturalistic inquiry of this type may have been the reason I opted to study science and chemistry and later, why the philosophy of science and the history of ideas became of such interest to me. It was not until I was at university that I more formally came across names like Copernicus, Kepler, Galileo and the person who could be described as the 'Father of modern science', Francis Bacon (1561–1626). Bacon articulated the 'inductive method' for science where conclusions are drawn from the accumulation of evidence, an approach that remains central to science today. Changes in method over succeeding centuries began to recognise the role of the scientist in formulating ideas and theories as distinct from relying on sense data alone caused a new account of science to appear, that of positivism. Karl Popper (1902–1994) proposed that science proceeds by 'conjectures' about nature that the scientist attempts to disprove. Laws and theories of science cannot be completely verified and shown to be true by observations and experiment, but they can be falsified and rejected. Conjectures are not mere opinion, because they can be tested, falsified and refuted by investigation. It was in 1962 that the next major change in our understanding of science occurred, with the publication of *The Structure of Scientific Revolutions* by the American philosopher of science, Thomas Kuhn (1922–1996). In this work, Kuhn challenged the idea that science was a linear process of mainly one method, with the theory of 'paradigm shift'. Kuhn proposed that there is revolutionary change at various times throughout the history of science where dominant views give way to fundamentally new understandings. For example, the corpuscular theory of matter, Newtonian mechanics, relativity and quantum mechanics. He pointed out that the change might not occur smoothly or all at once and required that the scientific community as a whole reaches agreement or consensus on the new characteristics. There is recognition by some practitioners that current thinking is inadequate or incommensurate in light of new experiment and evidence.

We can see a type of Hegelian or Marxist dialectic at work here, where the new emerges from the old and takes understanding to a different level while maintaining the old within its base traditions and practices.

Why some ideas 'ring true' to some and not others can be difficult to explain without a complete life profile of the people concerned. However there does seem to be a lineage or close fit between dialectic, inquiry and paradigm shift, or to put another way, between Hegel, Dewey and Kuhn that resonates with me. What needs to be added to whatever methodology is applied, is how, once evidence has been collected, the researchers make judgements on meaning, tentative of more conclusive. Feyerabend (1924–1994) proposed that science proceeds in a manner similar to other social investigations through hunches, intuitions and estimations, but such as approach, rather than being simply 'beyond method' (Feyerabend, 1975), could be seen as method in its own right. Both the social scientist and physical scientist must draw upon their experience, reading, discussion with colleagues and established protocols before deciding on the overall research methodology and the next steps, let alone interpreting results once obtained. Whether this approach involves what might be called hunches, or professional judgements, may be a moot point. It is not my view that thinking of social and physical sciences as having similar processes diminishes each, or enables one to colonise the other, but enables each to have a more complete view of the issues at hand. Each can be seen as a social practice with specific characteristics. Whether mathematics has a similar persona, we can investigate.

Mathematics as social practice

I am including mathematics under the broad heading of science here for the purposes of discussion. That is, when thinking about science, one often connects with our understanding of mathematics as well. This connection was unavoidable for me when, as a beginning mathematics and science teacher at a secondary school, my classes were evenly distributed between both subjects and across a number of year levels. My week therefore ranged between working with younger students who had very little experience of science and senior students who were grappling with difficult abstract ideas and procedures of mathematics and science. Given my reading of Dewey and philosophy generally, I was attracted to framing teaching as 'integrated knowledge and inquiry learning' and attempted to organise all classes as much as possible around various activity. This meant that my view of formal mathematics came about first and foremost through my observation of students and through my determination to establish learning by doing and making. That is, my appreciation of mathematics emerged from my actions rather than being told what school mathematics was all about. I knew from my background in chemistry that we pursue problems by using recognised methods or techniques, but at the same time, Feyerabend may be right in how we consider the meaning of what we are doing. For example, Wittgenstein (1958, p. 225) noted:

> One judges the length of a rod and can look for and find some method of judging it more exactly and more reliably. So, you say, what is judged here is independent of the method of judging it. What length *is* cannot be defined by the method of determining. To think like this is to make a mistake. What mistake? To say, 'The height of Mont Blanc depends on how one climbs it' would be queer.

Wittgenstein is making an important statement about mathematical certainty, or more to the point, uncertainty. There is a distinction being made between what we do and how we think about what we do. He goes on to say that we learn about the meaning of 'length' and 'determining' not by the words themselves, but by 'learning, among other things, what it *is* to determine length' (emphasis added). This can be taken as a severe criticism of school mathematics where stress can be placed on the technique of measuring length and area for example, rather than an understanding of length and area themselves. Wittgenstein (*ibid.*, p. 225) then makes a challenging statement for all educators, not only those involved with science and mathematics:

> Ask not: What goes on in us when we are certain that . . .? But: How is 'the certainty that this is the case' manifested in human action? While you have complete certainty about someone else's state of mind, still it is always merely subjective, not objective, certainty. These two words betoken a difference between language games.

We will come to a discussion of subjective and objective later in the book and what is meant here by 'language games'. For the moment, there is the epistemological consideration of how we come to understand through our actions in the world and indeed, whether we can trust our actions to provide accurate sense data of the world. Mathematics as it is conducted in schools needs to defend its position of mainly transmitting predetermined content and the certainty of problems with right/wrong answers, or whether it needs to open up ideas and see where they lead. This is a philosophical and ethical consideration of knowledge. Dweck (2012) and Boaler (2016) have approached this issue through the concept of 'growth mindset', whereby humans can develop their capabilities and achievements through activities that generate a love of learning and accomplishment. Boaler comments on the 'fixed mindsets' that many adults and students have about mathematics and how many are excluded from this category of human knowledge. At this stage, it seems a first hypothesis we could propose would involve a transformation of school mathematics from traditional rote methods of instruction, to immersion in dialectical processes of action and inquiry, not to produce answers with certainty, but to experience connections with knowledge and what the world has to offer.

Let me now make some brief comment about what Ernest (1998) calls the 'social constructivist' view of mathematics. By this, he means that the objects, theorems and procedures of mathematics are social constructions and therefore are fallible and can be constantly revised. Our human-derived descriptions of the patterns and relationships of the universe are just that, rather than being truths that have been discovered and are then taken as existing in their own right. That is, humans 'construct' their knowledge of mathematics from their experience and social practice in the same way as all other knowledge. There is an interesting question when we consider how constructivism generates new ideas when we have not had experience of the object involved, or to put another way, how do we construct our own abstract ideas, from nothing. What sense can we make for example of an abstract expression like $A \times B = C$? While not having seen or thought about this object before – although the symbols 'x' and '=' may have been encountered – we could immediately relate it to what we do know or have experienced, such as the alphabet, or a television programme and begin to make connections, particularly in discussions with others, the social. In this way, we are indeed constructing something from nothing. Those who doubt the social practice constructivist view of knowledge need to be able to propose where knowledge comes from generally, but particularly for children and students. I presume that knowledge that is not personally constructed, is transacted or taught, initially known by someone else who can then inform those who do not know. We must ask, how did those people come to know in the first place? These perplexities have somehow strengthened my view that mathematics and science fit within a broad understanding of human growth and do not sit outside where a different truth and cognition exist. In particular, they encourage me to involve students in this journey as well where we all set about engaging and listening to the world and becoming more human as we walk and talk together.

Knowledge and the humanities

I want to make some comment about the humanities in relation to the discussion of science and mathematics above. Arising from my understanding of American Pragmatism, I see all knowledge as integrated rather than discrete and human participation with all knowledge involving processes of inquiry and reflection. Allow me to repeat that I approach these questions from the perspective of epistemology and thinking rather than methodology and procedure. That is, what I am thinking about when fixing a puncture on my bicycle, as distinct from following the procedure of repair. On the surface, there appears to be an obvious difference between mathematics and science and the humanities regarding the type of data encountered. If we take the humanities as fields of knowledge that study societies and their cultures, then the data involved are more descriptive, reflective and speculative, rather than obviously empirical (Aronson and Laughter, 2016). At least on the surface. Generally including fields such as anthropology, archaeology, classics, history, linguistics and languages, law and politics, literature, performing and visual arts, philosophy and religion, we can see how practitioners will draw upon a range of data at particular times to inform the problem at hand, data that links the scientific and humanistic. Considering the type of glazing used will help date a clay pot for example found in an archaeological dig. Collecting stories from Elders about remedies they use for illness will indicate the historical trend of medical treatment in the community. Apart from the practical outcome of a better understanding of ourselves, our societies and cultures, a study of the humanities can be undertaken for its own inherent good and satisfaction (Hansen, 2014). Reading a historical novel on a cold winter's evening, learning another language, attending a play at the local theatre, discussing a contentious political issue with the next-door neighbour, all enhance human contact and enable us to have a better and warmer understanding of who we are and our place in the world. I would argue of course, that the sciences and mathematics should do exactly the same.

How do I conceive of the humanities and their impact? I well-remember being introduced to John Steinbeck's book, *The Grapes of Wrath*, when I was a secondary school student. It made a big impression on me. For some reason, I related to a group of farmers in Oklahoma (a place far away) who were struggling with economic depression, drought and troubles with the bank. At last they were forced to pack up and set out for California in the hope for better times including sunny days and picking oranges. I recalled stories from my own family as well that were passed on from generation to generation around the kitchen table involving hardship, being evicted from homes and mass unemployment. Steinbeck's writing of these times stirred something in me, including *Of Mice and Men* that we also read at school. I asked my mother to see if there were any more of his books in the public library and I remember she brought home *Cannery Row* and *Travels with Charley*. Steinbeck's description of travelling around the United States with his dog Charley not only told me much about that country, but inspired me to try and do the same in Australia, some day. All members of my family were great readers and I grew up in similar vein, being taken to all sorts of mysterious places and events in my imagination, as I read every night, mainly about the adventures of British children, school boys and empire. From these early experiences, I see reading as a central aspect of all knowledge domains and those who can convey ideas in writing and other media as having a crucial role to play. Take this passage from Hannah Arendt for example, where she is explaining her notion of *vita activa* in terms of work, labour and action (Arendt, 1958, p. 198):

> The *polis* properly speaking, is not the physical-state in its city location; it is the organisation of the people as it arises out of acting and speaking together and its true space lies between people living together for this purpose, no matter where they happen to be. 'Wherever you go you will be a *polis*.' These famous words became not merely the watchword of Greek colonisation, they expressed the conviction that action and speech create a space between the participants which can find its proper location almost anytime and anywhere.

We can categorise Arendt as working in the humanities, or in philosophy as a political theorist and a person who is writing for practical understanding, not so much for the pleasure of reading itself. In discussing the nature of *polis* (Greek: community of citizens, or city, city-state), she refers to the centrality of 'acting and speaking' and, in effect, the process of democracy being established between people every day as Dewey argued in terms of associated living and conjoint experience. Arendt (p. 199) goes on to note the significance of our explicit 'appearance' in the world such that 'the reality of the world is guaranteed by the presence of others, by its appearing to all, "for what appears to all, this we call Being".' These are difficult philosophical concepts where it could be said that she is drawing upon not only Dewey but Mead and his theories of society and subjectivity (to be discussed later). I did not come across Arendt until I had been teaching and lecturing for many years, which, I think, is both a weakness and a strength for education. My study, work and teaching in science and chemistry meant that I brought my science experience to bear on the humanities, rather than what I found is often the case, humanities teachers either ignoring science processes, or considering a smattering of science ideas that they come into contact with informally. My shift in emphasis from science to knowledge generally including philosophy and the history of ideas, prompted me to read more widely and to locate science in the broad sweep of culture, history and politics. Coming across Dewey by accident in the bookshop started this quest, taking me ultimately to people like Hannah Arendt and encouraging me to link ideas across what others may think as very disparate fields.

Over more recent years, my searching through the humanities has revealed a particular viewpoint called 'critical pedagogy' and a particular group of theorists who are very active in the field. My attention was drawn to this group when I read a book by Carr and Kemmis (1986) called *Becoming Critical: Education, Knowledge and Action Research*. This book impressed me greatly, not only because of its Australian co-author, Stephen Kemmis and its broadly historical and philosophical character, but because it introduced me to the German social theorist, Jurgen Habermas. I may have struggled with the ideas of Habermas, but the notion of 'becoming critical' in education struck a chord. Carr and Kemmis (p. 131) wrote about education as a 'critical social science' and stated, 'One of the central aims of critical theory has been to reassess the relationship between theory and practice in the light of the criticisms of the positivist and interpretive approaches to social science which have emerged over the last century.' These words jumped off the page and fitted entirely with my thoughts of there being no distinction between practice and theory in practice. I found that Habermas had written extensively on what he called the 'theory of communicative action', where he distinguished between the 'strategic action' of those who wanted to impose their will and 'communicative action' that relied on the reason of argument. I thought this to be an entirely democratic position to adopt. Habermas elevated human language and communication to centre stage in the process of learning and put forward a number of principles that would establish non-coercive interaction and understanding. My limited view of 'critical' involved a comprehensive rather than narrow consideration of issues so that analysis is all-sided and judgements defensible. I took it to mean a 'reflexive' view whereby our notions of knowledge change through practice and that we do not oppose this change process, the intellect becoming aware of itself through experience. From this reading, I traced the work of similar theorists who supported 'critical pedagogy', American writers such as Michael Apple, Joe Kincheloe, Henry Giroux and Peter McLaren. What I liked about all of these authors from the humanities was that they had a view of people and of knowledge that respected how each came together, in fact I could suggest, how each constitutes the other through the process of contradiction and sublation. I was able to relate their thinking to the practice of my classes whether as a secondary teacher or university lecturer and it appeared that my theorising of what it means to be human was consolidating.

We now turn to my connection with Paulo Freire. Like many other countries, Freire came to prominence in Australia during the 1970s through his work on literacy. He still has a strong influence today in adult literacy, reflecting I suppose where he began with villagers in the Brazilian countryside. Freire set up what he called 'culture/learning circles' to consider issues of importance to the village and how they could be solved. In other words, literacy begins with genuine human interest and the changing of social circumstances so that interest and meaning is enhanced. His approach within the culture/learning circles was dialogical, whereby initial discussion outlined main aspects of the issue, from which key themes were identified. These were called 'generative themes' because they clarified what was significant and produced new ways of thinking about the problem. Finally, judgements were made about the best way to proceed and results were evaluated as action proceeded. Cycles of investigation continued in this manner until a satisfactory outcome was reached. Freire was clear that literacy involved both action and reflection in cycles of exploration and not the mere adoption of so-called known truths and procedures, in linear fashion. He famously declared (Freire, 1972b, p. 104):

> Scientific revolutionary humanism cannot, in the name of revolution, treat the oppressed as objects to be analysed and (based on that analysis) presented with prescriptions for behaviour. To do so would be to fall into one of the myths of the oppressor ideology: the *absolutizing of ignorance*. This myth implies the existence of someone who decrees the ignorance of someone else.

Statements like this fused very strongly with what I saw as my evolving role as a teacher and educator obligated, like Freire, to social justice. I did not accept for example that my students in mathematics and science were ignorant regardless of their mainly working class background, in fact, quite the reverse. I understood that it was my role to somehow raise the big ideas of these areas and connect them with the cultural and historical experience of those sitting in class. Social and intellectual connection was the basis of teaching I thought. I needed to begin with the practice and experience of families and students in recognition of the vast intellectual reservoir at their disposal and then, through cycles of learning circle investigations, expose the beauty and sovereignty of knowledge as means of changing the world. At this stage, I was more than sanguine in linking my fascination with knowledge from mathematics, sciences and the humanities and reflecting on where it comes from. I was being guided by the general ideas I was reading from Dewey, Freire and others about how the theorising of personal and community practice occurs. Of course, education policy, the curriculum, testing and many colleagues had vastly different ideas to mine.

Extending practice to praxis

I was very fortunate to begin my employment as a university lecturer in a faculty that had social and educational practice as its guiding principle for teacher education. There were two main reasons for this somewhat unusual policy position. First, the university was located in a working class region of Melbourne and an emphasis on practice established close links with the culture and working experience of local communities. This was in accord with a key establishment aim of the university to respect and serve working people of the region many of whom may not have attended university before and enabled programs to be informed by local knowledge and aspiration. Second, social and educational practice connected nicely with the interests of a number of lecturers who, at the time, were influenced by work of the British sociologist Anthony Giddens and the ground-breaking studies by Australian academics Stephen Kemmis, Robin McTaggart and colleagues regarding Action Research. Giddens became very well known in the 1980s for his attempts at reviewing and strengthening principles of sociology, out of which he proposed his 'theory of structuration' involving the interrelation between structure and agency (Giddens, 1984, 1993). Kemmis and McTaggart (1988) published their initial booklet on action research in 1984 and, although only a small volume, encouraged many teachers and other practitioners around the world to investigate their own practice for improvement (for a history of action research, see McTaggart, 1991). There was thus a strong theoretical basis at the university for socially-just, practice-based teacher education programs and the establishment of democratic partnership arrangements between the university, schools and other community organisations. I was very pleased to become a part of that movement.

Any new or revised programme in education must of course answer the question and often criticism, 'does it work?' I am not surprised to hear this, but it surely should be directed at current, conservative approaches, as well as new, progressive proposals. One of the problems in answering this question is the different student outcomes intended and whether such outcomes can be 'measured' by traditional techniques such as paper and pencil examination. As mentioned above in relation to mathematics, it is one thing to improve test results but keep our understanding of mathematics the same, contrasted with an improved epistemological understanding, by reconceptualising the very nature of mathematics as a field of knowledge. In our case, we introduced a professional portfolio for our graduating groups of teachers whereby they documented and reflected on their work, not only for their final year of study, but throughout their entire course. Discussions or interviews were then held at the end of final year around a set of professional questions that required candidates to draw from evidence contained in their portfolios. Questions were of the type:

- What has interested you most throughout your course?
- What were your most significant challenges faced this year?
- How do you prepare for diverse classrooms?
- Which theorists and books have you found of most assistance?
- How do you approach the issue of equity and social justice in education?
- What are your plans for professional learning next year?
- How would you like to see yourself as a teacher in 5 years' time?

I was very impressed with the quality of discussion at our portfolio interviews. Candidates presented and spoke in a professional manner and, most importantly, were able to defend their comments with a range of portfolio evidence. In talking about equity, diversity and social justice strategies for example, candidates referred to flexible lesson planning, advice from mentor teachers, contact with families, discussion with lecturers and academic writing from their university units of study. I took this to be the end result of a 4-year course that emphasised social and educational experience and practice, reflection on teaching practice, connections with broader understandings through literature and discussion and commitment to children and the education profession itself. Before I had come across Peirce mentioned above and in reporting to colleagues after portfolio interviews, I used the term 'clarity of understanding' that I was confident candidates demonstrated. I am comfortable in suggesting that our graduates were working through the three levels of clarity suggested by Peirce, that is familiarity with the topic or object, discussion of the topic under different circumstances and observation of the effects of the concept in a comprehensive manner. I am using description here as a valid and credible measure in support of practice-based learning, an intellectual process of connecting actions, thinking, reflections and discussions for personal awareness of what the world is all about.

In the mid-2000s, our teacher education team decided to strengthen our approach to learning through practice and partnership by incorporating the idea of 'praxis' as a major guiding principle (see for example early work by Cherednichenko and Kruger, 2005/2006). By this time, there were a number of staff in the faculty who were enthusiastic about Freire and were keen to pursue his notion of praxis, defined as the coming together of theory and practice for change and improvement of social and educational conditions. Others have embraced this approach as well in various ways, including Marx, Lukacs and Gramsci who supported a 'philosophy of praxis' (Feenberg, 2014), meaning a broad understanding of society that encompasses the relationship between the bourgeoisie and proletariat. Kemmis and Smith (2008, p. 4) summarise a range of views when they write:

> *Praxis* means different things in different intellectual and cultural traditions. In some European traditions for example, *praxis* is understood as any social action undertaken in the knowledge that one's actions affect the well-being and interests of others. In other traditions, notably Marxian traditions, *praxis* is the kind of action that makes transformations in the social world. In this book, we regard *praxis* as a kind of enlightened and 'elevated' action.

Our thinking about praxis provided an enhanced theoretical frame for partnership work with schools and other organisations and enabled us to develop our own interpretation to suit local conditions. Some colleagues referred to Greek philosophy and noted the idea of *phronesis* as the intention of citizens to live well and *praxis* as the action arising to put this intention into effect. Such a view located teacher education and indeed education itself, in a long history of philosophical debate and suggested 'education as philosophy of practice'. According to the subtitle of his book *Democracy and Education*, Dewey himself regarded education as philosophy, as distinct from philosophy of education, but did not seem to position this in relation to Greek thought. A literature has developed over recent years on praxis, education and teacher education including our own writing (Arnold *et al.*, 2012; Hooley, 2015) and authors such as Kemmis and Smith (2008) and Kinsella and Pitman (2012), but this has not been mainstreamed as yet. University programs are of course restricted by budget, the short time they have for teaching of study units, meeting university requirements of Academic Board and external approval bodies. However praxis in action can be seen in programme design, partnerships with organisations and the portfolio process and interview noted above. In setting up daily class organisation for example, a Freireian approach can be adopted involving semi-autonomous learning circle negotiation of a topic of interest, gathering of data, identification of generative themes and consensus on options and action arising. This process of necessity is practice-based, cultural and democratic and is the action arising from living well. Another example could involve the introduction of 'capstone' classes for final year candidates (see Chapter 11). Using the analogy of the central stone in an archway or bridge that holds the rest of the structure together, capstone programs are intended to bring key features of a course of study together in an integrated and innovative manner. They may encourage more creative and non-traditional methodologies so that candidates realise the range and diversity of data available for knowledge construction. At this stage, long-term research and evaluation projects of praxis-oriented education and teacher education have not been made, but the strong theoretical base, documentation and small-scale research projects that have been undertaken are strongly supportive of the practice/partnership/praxis triumvirate as steps forward in quality education and epistemological progress.

There is a significant historical and philosophical question that arises from our discussion of various knowledge disciplines and practice/partnership/praxis: what does this tell us about human well-being, indeed the quest for that most elusive and unclear human disposition, happiness itself (Grant, 2012)? We have not only been thinking here about the policy and practicalities of schooling, learning and teaching, but education as philosophy of practice that constitutes social life, our very being. This is an extended view of education's role in establishing social justice for all citizens and eliminating or at least mitigating the ravages of social division. In this regard, Dewey (1897, p. 77) commented, 'I believe that education is the fundamental method of social progress and reform', and that in defining the nature of education:

> I believe that all education proceeds by the participation of the individual in the social consciousness of the race. This process begins unconsciously almost at birth, and is continually shaping the individual's powers, saturating his consciousness, forming his habits, training his ideas, and arousing his feelings and emotions. Through this unconscious education the individual gradually comes to share in the intellectual and moral resources which humanity has succeeded in getting together. He becomes an inheritor of the funded capital of civilization. The most formal and technical education in the world cannot safely depart from this general process. It can only organize it; or differentiate it in some particular direction.

As we see from these extracts, Dewey had a very broad concept of education as a central characteristic of being human, of how we live and interact with the social and physical worlds of our experience. That being so, we also need to consider where our understanding of pragmatism, inquiry, practice and praxis takes us, a naturalistic philosophy of existence and what it means to be human. This reflective issue was expressed earlier in the Preface where I noted that this book draws upon Greek and European philosophy that asked, 'How should we live?' and European Enlightenment that considered, 'What can we know?' to question today, 'What does it mean to experience mind, to act, think, know and create ethically?' This is the question of human subjectivity, of humanity itself, of what it means to be. I have approached this task from the point of view of epistemology, or how I understand knowledge and learning and, in particular, implications for all citizens when living under the economic dictates of neoliberal capitalism. When I think of what it means to be human, the following characteristics come to mind:

- Autonomous
- Conscious
- Creative
- Democratic
- Knowledgeable

- Sensuous
- Sentient
- Sexual
- Social
- Subjective

These characteristics are extremely difficult to define and extend across issues such as class, race, gender and the like. That is, these are the great unifying features that bring all people together regardless of economic, cultural and historical difference. Based on these features, we can then consider the conditions by which such beings can 'experience mind, to act think, know and create ethically'. I see this as the formation of an awareness of subjectivity that comes about as we act in and with the world, of our personal capabilities, knowledge and judgements about what is real. In relation to institutions that we have created such as families, schools, churches and the law, I suggest at this point of our discussion that we consider the notion of 'intersubjective praxis' as a way of summarising the characteristics above and of thinking about 'how we experience mind'. For families, schools, churches and the law, *praxis* is taken to be the action, practice and transformations of *phronesis* that enable a humble and public life, while *intersubjective* means the collective understandings that are constructed as we pursue the wholeness of mutual human interest and happiness. A philosophy of practice of this magnitude is emancipatory taking us well beyond dominant and aggressive economic imperatives.

Case 2. Thinking about the clarinet

Generally speaking, Alice didn't mind her older brother Michael and his circle of friends, although they could be annoying sometimes. They seemed to be getting involved in the local surf club these days and were spending a lot of time there. This suited Alice, as she had just begun secondary school and wanted as much peace and quiet as possible, so she could concentrate on her studies; there was so much to be done. She liked reading and could curl up with a book for hours, being drawn into the world of imaginary characters that was very different to her own. Apart from swimming and surfing, Michael had a passion for music and was attempting to learn the trumpet. This drove everyone in the house to distraction as he hadn't made a lot of progress so far, but mum and dad said that give him time and the loud noises emanating from his room would someday turn into sweet cool music. They hoped. When he wasn't huffing and puffing, Michael would play a lot of videos and pop music on his laptop, sometimes singing along equally loudly. One day, dad brought home a number of old CDs that had been on sale and said to Michael that he might be interested in the different styles of music that they featured. Pretty soon, Michael was experimenting with what he thought about music although he proclaimed that 'classical' was not for him. On the other hand, he was attracted to jazz for a while, but then started playing over and over, what he called 'big band swing'. Apparently, this involved bands of about twelve members who played all the instruments, brass, strings, drums, piano, a style popular many years ago. Alice couldn't help hearing what was going on in the room next door and found herself tapping her fingers and feet to someone Michael called Benny Goodman. Benny was a clarinet player and leader of a famous big band from America at the time. Quite unexpectedly, Alice became engrossed with this music and she didn't know why, except perhaps it had a beat she could hear and a rhythm she could feel. When listening, her mind drifted to other places, just like when reading her favourite books. Mum and dad always encouraged their children to try new things, so when Alice said one night that she would like to learn to play the clarinet the same as Benny Goodman, they were a little surprised, but pleased. Like Michael's trumpet, the clarinet was expensive, but Alice had a sense of excitement when, trembling slightly, she put it to her lips and blew for the first time. Nothing happened. She tried again, but no matter how hard she blew it was impossible to make a sound. Alice was devastated and near tears, but Michael said, to his credit, 'Not to worry, the same happened to me, just keep trying.' Mum asked someone at work and they said to try a different reed that might require less effort. Success. When Alice blew and got that first note, it was the best ever, sweet and cool to her ears, Benny himself would have been proud. As she fingered the keys and experimented with different notes, Alice sensed something important had happened. She wasn't sure why the beat, rhythm, harmony and sound itself caused her to feel like a different person, indeed she didn't even know what those musical things were. But she knew that she was about to find out.

3

INTERSUBJECTIVE PRAXIS

Thought erupts from somewhere
responsive heart of emotion
perhaps a starry starry night
where Vincent's troubled sense
at once breaks free of tortured life
clasping emergent waves of awareness.

To propose and discuss education as an emancipatory philosophy of practice is a startling, indeed revolutionary claim, meaning that we need to have a shared understanding of those topographies of philosophy that lie behind this idea, the complex notions of ontology and epistemology. My introduction to the concept of ontology came from my reading of Giddens, although I thought that his view was more sociological than philosophical. In contrast and somewhat later, a colleague and myself wrote 'In simple terms, ontology is the philosophical inquiry into "reality, nature of existence, or being"' (Mills, Durepos and Wiebe, 2009, p. 630, cited in Samaji and Hooley, 2015). An ontological view then asks questions about human being, what is real and where does reality come from. Samaji and Hooley went on to describe epistemology as the philosophical questioning of knowledge, the assumptions upon which it is based and therefore questioning what we 'do know' and 'can know' (Allison and Pomeroy, 2000, p. 13). From an epistemological perspective, questions will be asked about the nature of knowledge, where knowledge comes from and how learning interactions with the social and physical worlds occur. It is doubtful however whether ontology and epistemology can be separated in practice after many centuries of evolution and it may be useful to consider both existing dialectically in a continuum of practice, what comes first, the chicken or the egg? My worry about taking a purely ontological view was that explanation of how we came to be the way we are today, still remained. If not, then the way we are has always been, from the beginning, taking us down a theological path. When educators and teachers are working with students, there is a need for epistemology to assist what we do in practice, something that ontology cannot.

How does placing epistemology first, so to speak, work in practice? A way of thinking about this question was provided by Piaget through the concept of 'structure'. Based on the idea of contradiction and the dialectic of Hegel, Piaget theorised human thought as proceeding from a state of 'assimilation', whereby new information is incorporated into current cognitive schema without changing that framework, to 'accommodation', that allows for the changing of existing schema in relation to the new experience and then 'equilibration', where a new level of understanding is reached to the satisfaction of the human organism. A dialectic exists between these processes with each continuing to exist within the other. Piaget's schema or structures were not easy to define but, in quoting the view of Levi-Strauss that structures belong 'to a system of conceptual schemes somewhere midway between infrastructures and conscious systems of conduct or ideology', Piaget (1971, p. 138) went on to say:

> But we, if asked to 'locate' these structures, would carry Levi-Strauss' suggestion one step further; we would assign them a place somewhere midway between the nervous system and conscious behaviour because, to adapt his locution, psychology is first of all a biology.

It is clear from these extracts, that Piaget and Levi-Strauss were grappling with what it means to be human and their thinking on structure had taken them to a point where there is a dialectic between ontology and epistemology. Theorising something like assimilation, accommodation, equilibration can be described psychologically, but the process also needs to be taking place biologically, or how humans are. In this way, humans actively construct new structures of understanding for themselves, rather than merely accept what they are told in a linear fashion. Critics of constructivism as a philosophical view of human knowledge (see Roth, 2011), argue that it is difficult to see how more complex structures occur from simpler ones and how new ideas can occur when there is no previous experience on which to build. Chemists and physicists are quite familiar with the creation of new more complex entities than what existed previously, such as with the products of chemical reactions and with the formation of new galactic substances. New objects of thought that reach a higher level of equilibration are entirely feasible. When undertaking a new experience, for example when contemplating a plate of food that has not been seen or tasted before, the person brings their sensory perceptions into play immediately relating what they know with what they do not. Initial connections made may not be totally accurate, but will be refined as the eating experience proceeds, in effect, constructing a new understanding from nothing, except the sum total of culture that guides all action. We can think of this process as a continuing becoming, as structures are always transformed into something else and the human subject becomes anew.

Before we go on, let us pause to briefly summarise some of the key ideas that we have encountered so far and on which we will base our next consideration. Table 3.1 contrasts three different approaches to philosophy, namely empiricism, rationalism and pragmatism, that overlap but allow us to envisage how humans exist in different ways. I have included what is generally known as British Empiricism and referenced the writing of the British philosopher John Locke (1632–1704) and the Scottish philosopher David Hume (1711–1776) to indicate connections between the three schools.

A summary table does not do justice to philosophy of course and the inspirational work of those who have and continue to spend lifetimes attempting to clarify our humanness. But one or two features are apparent. In the first instance, agreed and contested characteristics of being human are categorised although the emphasis and interconnectedness given to each differs markedly. Second, the social and political context within which ideas are formulated cannot be denied, with the European and American political, cultural and industrial experience having a strong influence. Today, for example, one could argue that our views are being fashioned by digital acquaintance as we may think of mind from a computer, information-processing perspective. Third, the table shows that we still operate from a descriptive model that has not changed dramatically together with the questions that we ask over the past 350 years, despite developments with neuroscience, brain imaging and the like. Finally, there are serious implications for

TABLE 3.1 Contrast between major philosophies

	Empiricism	*Rationalism*	*Pragmatism*
Philosophy Positioning	British Empiricism Hume, Locke	Continental Rationalism Kant. Descartes	American and European Socioculturalism Hegel, Marx
Theorists Education	Skinner, Watson, Gagne	Piaget, Chomsky	Peirce, James, Dewey. Mead, Addams, Freire, Vygotsky
Ontology Being	Provided by being human	Interaction between biology and consciousness	Accords with the natural processes of the universe
Epistemology Knowledge	Concepts and knowledge gained through sensory experience	Concepts and knowledge gained primarily through reason	Concepts and knowledge gained through experience of the effects of objects
Pedagogy Practice	Changing behaviours through instruction	Action and reflection through experience	Social and physical integrated knowledge and inquiry learning

how schooling is perceived and organised with each underlying philosophy supporting totally different curriculum, pedagogy and assessment. One major aspect that bears directly on these features is how the emerging subjectivity of each child is understood, perceived and supported. At a later point in our discussion, I will attempt to relate subjectivity to consciousness, but for now, let us think about personal subjectivity as our totality of feelings and interpretations of the world that guide our social engagement in a two-way, indeed dialectical manner. I tend to see this sensorium as the human brain because I'm not sure of what else there is to undertake this task, but my viewpoint can be taken as being too narrow and cognitive. To expand on this point therefore, we return to pragmatism and the connections between action and thought.

Mead, subjectivity and intersubjectivity

I had come across the well-known book about George Herbert Mead entitled *Mind, Self and Society* (Morris, 1934/1962), but had not paid a lot of attention until I attended a seminar by one of my university colleagues. As part of his presentation, he outlined a process of 'internal conversation' that Mead proposed as occurring when we were involved with others, or more accurately, a 'generalised other'. These concepts seemed to connect with my emerging interest in human consciousness and how reflection on experience occurs. I decided that Mead might be more useful for my teaching than I had previously thought and required investigation; this quickly took me to the work of Gert Biesta, mentioned previously. George Herbert Mead (1863–1931) was a friend of John Dewey when at the University of Chicago and they appeared to have a close professional and personal relationship. I understood that consciousness was a continuing problem for philosophy that had not been resolved over the centuries. For example, there was a view that consciousness was separate and individual, arising from we know not where, but this was taken to explain why some people were the way they were, different from other people walking down the street. If my view of materialism meant there is nothing else except matter and energy, then how we think is based on the human characteristics outlined above and evolves as a result of human action in all its forms. I was comfortable with the notion of interchangeability between consciousness and subjectivity, both unclear terms attempting to describe where ideas, thought and guidance for practice comes from. This was an important point to reach in my understanding, as it sets up the conditions for equity in schools. That is, we have a view that in every class, there are groups of students and teachers who are sentient, sensory beings grappling with the complexities of their existence. All begin from the starting point of human characteristics and the consciousness and subjectivity they embody, changing as new experience is accumulated. It is clear that schooling systems around the world have not resolved ways of pursuing this historic journey in the interests of all.

My growing interest with consciousness and subjectivity raised serious questions about our humanness such as kindness, sacrifice, generosity, selflessness, courage and the like. Where do these inherent qualities – and their opposite – come from and how are decisions made to exercise them? What is the basis of ethical life? Biesta and Trohler (2008, p. 5) provide a clue when they write about the derivation of human meaning from our reaction to objects:

> Mead's point here is that objects do not have any meaning as such; they do not have an 'objective' meaning. Their meaning lies in what they mean *to us* and this is to be found in how we respond to them. To '*get*' the meaning of an object is, in other words, not a process of discovery, but a process of *creation*. This is not to say that any meaning will do. Both with respect to physical and social objects, some responses will be more adequate, more appropriate, or more functional than others.

I take this insight to mean that to understand an object or idea like 'coat', we need to come into contact with 'coat' in some way so that, rather than imposing our initial response on the object, the object will engage us and our thoughts fluidly. In this way, Mead seems to be supporting the notion of dialectic, with the object and ourselves existing in relation to each other. How this process begins cannot be determined in the hustle and bustle of everyday life, but Mead argues that it is the 'social act' that is the precursor of thought. This concept is one of the central issues of the philosophy of education and of pragmatism, although it is given little courtesy in the educational literature today. Most significantly, Mead located individual social acts within the broader group where individual acts go 'beyond' and in turn 'implicate' the entire group or community. It is worth quoting Mead's understanding of the social act in full (Morris, 1934/1962, p. 7, Footnote 7):

> A social act may be defined as one in which the occasion or stimulus which sets free an impulse is found in the character or conduct of a living form that belongs to the proper environment of the living form whose impulse it is. I wish however to restrict the social act to the class of acts which involve the co-operation of more than one individual and whose object as defined by the act, in the sense of Bergson,[1] is a social object. I mean by a social object one that answers to all the parts of the complex act, though these parts are found in the conduct of different individuals. The objective of the acts is then found in the life process of the group, not in those separate individuals alone. ([1]Henri-Louis Bergson 1859–1941, French philosopher and Nobel Prize winner 1927; note added.)

For a person who supports Dewey's approach in particular, I find it difficult to think of social act as anything else than a cyclic, dynamic and perpetual interaction between objects and thought, in the same way as I cannot isolate practice and theory. I accept therefore the object 'coat' only becoming 'real' to me when it comes into relationship with me through use. Through using 'coat' to keep warm and to keep dry from the rain, I understand 'coat' in a way that was not possible before. Mead makes the conceptual leap for epistemology when he defines 'social act' as taking place socially, with group understandings being forged and refined. These acts then constitute the 'life process' of communities, as distinct from the experience of each individual. Mead's philosophy was distinctive in seeing human being as inherently social, from the outside in so to speak, rather than being determined from the individual and each separate consciousness. He gives added detail to this process through describing what he calls the 'generalised other', or the 'unity of self' that arises from the organised community or social group (Morris, 1934/1962, p. 154). My way of thinking about this concept could involve what happens when we walk into a train carriage in the morning on the way to work. As we walk in and try to find a seat, we note the diversity of passengers present, perhaps office staff, construction workers, middle managers in suits and a selection of school and university students, people of all ages and backgrounds, some reading, some sleeping, some looking vaguely out the windows wondering or apprehensive at what the day might bring. We probably note familiar faces and new additions such as a woman in a different cultural dress, or a different school uniform. We instantly become a part of that group or community, drawn together by a common assignment of meeting the requirements of the day and a hope that it will go smoothly. As I observe two shoppers talking loudly and wonder whether I should indicate disapproval, I decide not to bother as they might stop in the next couple of minutes. My location with this group of train travellers is as an individual with my own aspirations, but I relate to them as a group, as a 'generalised other', with an understanding of the culture of train travelling in the morning and what might be going through the collective mind. I have a spontaneous view of what is appropriate and how to act. Mead's notion of 'internal conversation' within the context of 'generalised other' seems reasonable and important for teachers in their complicated and vibrant classrooms.

Based on his emphasis of the social, Biesta notes that Mead developed a social conception of education that conceives of subjectivity and thereby learning taking place 'between' human beings rather than emerging individually. He points out (Biesta, 1999, pp. 482–483):

> This philosophy of intersubjectivity not only accounts for the emergence of reflective consciousness in the process of social interaction. It also entails a redefinition of the process of social interaction itself, in that it does not conceive of this process as constituted by the activities of self-conscious subjects, but rather sees it as a matrix of coordinated action, a matrix, moreover, in which meaning is not simply reproduced or transmitted, but actually created.

These are descriptive concepts or models of what might be, but they paint a very active picture of being human rather than a passive, merely receptive role. The concept of reflective consciousness is challenging giving rise to a process that is not only reflective of personal action, but is reflective of itself. However this is necessary if consciousness and/or intersubjectivity is able to create its own meaning. I will take such creation as a process of construction discussed previously, whereby new cognitive structures of some type are formed from the old. Mead also detailed social interaction as involving 'gestures' of all parties including physical and verbal signs that needed to be observed and assessed, part of the 'matrix of co-ordinated action' noted by Biesta. The issue of gesture, including that of language, will be taken up later. There is the added question of the notion of 'meaning' and how this is derived within the human organism. This is a similar question I think to that raised above regarding kindness, generosity and other human traits that seem to emerge from somewhere that is almost impossible to explain. Pragmatism would propose that such features come from action, but that does not explain the 'feeling' let alone 'meaning' of kindness when it occurs. It may be that there is no 'meaning' in the universe at all, absolute or tentative, only the meanderings of human prospect and belief. Such feelings may be characteristic of the various arrangements of carbon, other matter and energies that constitute the human organism and are there to be accepted not explained. We do need to have some appreciation of this concept however if my original question of 'what does it mean to experience mind?' is to be pursued.

Like Dewey and the Pragmatists, Mead lived during an exciting time regarding the development of modern science, including the theories of Marx, Darwin, Freud and Einstein. As well as providing logical explanation for what is, science raised a number of problems for philosophy and religion including the view that scientific understanding was the highest form of knowledge and that scientific law therefore would dominate other views that formed the basis of human values, ethics and aesthetics. Was modern science arguing that the laws of the universe and of nature were predetermining a certain future, or through interaction with the environment, an unpredictable evolution? Joas takes up this question and how Mead was compelled to consider Einstein's theory of relativity and space-time from the point of view of human subjectivity. According to Joas, Mead suggested a 'common world' of humanity where different points of view are able to be held simultaneously and relatively in a temporal and spatial sense and, by so doing, '*Mead is laying the corner-stone of an intersubjectivist theory of the consciousness of time*' (Joas, 1997, p. 188, emphasis included in text). Time is a significant concept for science and philosophy and indeed for action itself and whether we think of ourselves as existing in a linear pathway of past, present and future, or whether we exist in a present, 'in a praxis taking place in a present, in an *intersubjective praxis taking place in a present*' (Joas, 1997, p. 192, emphasis included in text). In this way, Mead is taking account of the historical changes taking place within science and how they relate to his notions of knowledge and meaning arising from social interaction, between rather than within and ultimately, how they lead to his social conception of education.

In the previous chapter, I opened up preliminary discussion on these issues by suggesting that we consider the notion of 'intersubjective praxis' at that moment in our reading as a way of thinking about how mind is experienced. I took *intersubjective* to mean the collective understandings that are constructed as we pursue the wholeness of mutual human interest and happiness and *praxis* to be the action, practice and transformations that enable a humble and public life. I think it is necessary to link intersubjectivity and praxis for two main reasons. First, our concern for the significance, quality, satisfaction and continuance of human life means a philosophy of living that must be lived, day-by-day, not only causally explained. Members of the citizenry will approach this question from a religious and metaphysical position, while others, from a secular and scientific position, although the two are not exclusive of each other. While questions of meaning and telos remain unclear and are subject to disagreement, we need a flexible framework of beliefs, values and ethics that guide how we act towards and with one another so that human life as we understand it, is meaningful. This requires an evolving appreciation and consensus regarding our mutual humanness, consciousness, sociality, subjectivity and intersubjectivity. Second, these characteristics need to be enacted if they are to be achieved, in praxis, where our thinking and action constantly interweave dialectically, each existing in the other and forming new concepts and new social acts. It is difficult to separate thought from act, or indeed the new thought and new act that are constantly creating subjectivity and intersubjectivity between humans. If I could slightly amend the quote from Joas above and to extend Mead, I think we are 'laying the corner stone of a philosophy of intersubjective praxis necessary for peace and harmony of the human mind'. This takes us to the question of the nature of sociality.

Sociality, intersubjective praxis and social cohesion

I have never been happy and consequently it is a word that I do not use. This is of little concern to me for two main reasons. In the first instance, happiness is very difficult to define and it is therefore difficult to agree to something that is unclear, confusing and seemingly out of reach. Usually other words such as contentment, satisfaction, pleasure, well-being are used in its stead. In the second instance, if it is to mean anything, then happiness must be considered as a near-permanent state of mind and according to Mead and the 'generalised other' concept, requires connections with communities of similar understanding to exist. Given that the vast majority of the world's people are consumed by war, aggression, discrimination, poverty, hunger and ill health, it is not likely that happiness will be a common 'generalised' condition. I have had fleeting moments of what could be described as delight, enjoyment, perhaps contentment, but these have been few and far between and disappear quickly into the void: being embraced by the ocean's changing personality, astounded by the stars at night, a friend's glance of recognition, tears when understanding a poem, a tender kiss good night. Whether these are moments of happiness I don't know, but they all engender a 'feeling' that is different, perhaps a sense of calmness, serenity, of connecting with something outside of one's self, beyond. While it is possible to explain other feelings such as pain, fear, surprise, embarrassment in terms of the release of chemicals in the body, we tend to seek other explanations when feelings mentioned previously such as happiness, kindness, generosity, love, gladness are involved.

Erich Fromm, a member of the Frankfurt School of social theorists (Jeffries, 2016), was a psychologist who distinguished himself from Freud in proposing that economic and political factors rather than innate drives constitute the basis of human conduct. He wrote that the deepest and most profound need of humans was to 'overcome separateness', to leave the 'prison of aloneness' (Fromm, 1956/2006, p. 9) and that what we call love is a means of establishing attachment. He went on to nominate the central aspects common to all forms of human love as 'care, responsibility, respect and knowledge' (p. 24). We can think of care as being active concern for that which we love, responsibility as being able to respond to and be responsible for others, respect as concern that the other person should grow and unfold as they would like and that knowledge involves knowing the other person when I transcend concern for myself and see others on their terms. Fromm comments that we need to know ourselves, but in a broader and more significant sense, the nature of humanity and its purpose: 'The further we reach into the depth of our being, or someone else's being, the more the goal of knowledge eludes us' (p. 27). It is here that complications of 'being in the world' arise.

Mead used the concept of 'sociality' as understanding 'being in the world'. As outlined in his series of essays and notes *Philosophy of the Present* (Mead, 1932/2002) and *The Philosophy of the Act* (Mead, 1938/1972), Mead adopts a dialectical and social approach to the emergence of ideas. According to Mead, sociality 'is the situation in which the novel event is in both the old order and the new which its advent heralds. Sociality is the capacity of being several things at once' (p. 57). It can be seen as having two modes. There is a process of adjustment whereby the organism brings an emergent event unto itself and the process by which the organism is capable of existing in a number of systems at once. Mead famously used the analogy of the bee existing in relation to flowers, other bees, animals, humans, in the same way that humans exist in relation to other humans, bees, trees, history at any one time. Pragmatism needed to theorise how the new or novel occurred from the present and similar to Piaget's structures and indeed Darwin's evolutionary process, propose change and transformation as the natural order of things. This intersubjective praxis approach to human existence provides a basis for human autonomy and emancipation, where sociality and the social act enable all humans to construct their own understanding of what is real through continuing interaction, communication and reflection with others. Authority shifts from external to internal locations and brings all capability into alignment over time despite the influence of political power and dominance. Dialectics, sociality and the social act can help position citizens regardless of background in appreciating their own existence and open up avenues for becoming aware of the 'experience of mind', but within that, describing such 'feelings' may still be vague and intangible. Table 3.2 compares sociality and sociology regarding understandings of subjectivity and praxis.

TABLE 3.2 Comparison of sociality, sociology

	Sociality *Associations of participants*	*Sociology* *Studies of societies*
Subjectivity	Awareness of thinking, knowledge through social acts	Collection of personal experiences, culture, beliefs
Intersubjectivity	Understanding through interaction between people, nature; being in two states at once	Relationships formed between people
Praxis	Ethical action for well-being and public good	Ethical action for well-being and public good
Intersubjective Praxis	Interactions between people that arise through ethical action for well-being and the public good	Relationships formed between people through ethical action for well-being and the public good

At this stage, I want to comment on the difference I see between 'sociality' and 'sociology', especially as that relates to education. We can begin with Giddens (1984, p. 345), who comments that 'All human beings are knowledgeable actors. "Objectivism" fails to appreciate the complexity of social action produced by actors operating with knowledge and understanding as part of their consciousness.' Here, Giddens connects actors, that is humans who act, with their conscious (and I presume unconscious) knowledge that are in constant relationship, each forming the other. It places the definition of knowledge at centre stage, where it comes from and how new novel ideas are created. This fits very nicely with Mead's notion of a 'philosophy of the present' such that the present draws upon the past and looks to the future. Biesta (1998) writes about Mead's understanding of the social in this way:

> First of all I want to follow Mead in the characterization of his own position as a social conception of education. This is not a sociological conception, in that it does not conceive of education from the point of view of the demands of the society in which the child is entering but rather wants to recognize both the child and society at once. The situation of education is not a situation where the child stands opposite to society, but one where the child is a participant from the very first day of its life onwards. When Mead stresses, therefore, that education comes back to producing a social situation, it is not to conceive of education as a means toward some end outside of education, but rather to confirm that no educator can produce a social situation on his or her own. Education only exists in the interaction between the educator and the student/child, it only exists 'in communication.'

Making the distinction between social and sociology is a significant issue for education, where this view is usually confused. In general terms, I see modern sociology as describing the situation of formal education and how this might impact on teachers, families and the child, whereas social understanding is concerned with action, how teachers, families and the child act within educational situations to pursue interests. Seeing all participants as social actors within the lived, negotiated experience of education and being 'in communication' produces a very different concept of formal schooling and its passive one-way transmission of predetermined information and occasionally, knowledge production. Joas takes up the concept of Mead's sociality in depth that takes into account the understandings of modern science and conceives of a nature that is not independent of humanity, but 'is a perspective of a nature which is unfolding itself through the universalisation of action and of cognition' (Joas, 1997, p. 186). Again, we are brought to the position where our thinking of education, schooling, practice and praxis, must consider humanity itself, our place within it and what that place might be in the extended future. Will sociality comprised of 'knowledgeable actors' create a more democratic, humane and ethical human existence?

I doubt whether what we see around us is the pinnacle of human progress, that we cannot expect to achieve much more. When Kepler and Galileo gave us a new understanding of the universe, they gave us a new understanding of ourselves. When Newton described his gravitational laws and 'action at a distance', he provided a different perspective on body and mind to that proposed by Descartes. When Einstein theorised a relational rather than deterministic universe, he provided a new concept of being human. All of these events are concerned with breakthroughs in knowledge and the construction of new paradigms of subjectivity for our ideation and continuance. What I think happens each time we come across a new idea, or for some reason, suddenly realise that we know something that we were struggling to understand, is that we 'feel' a new consciousness or awareness of our own capability to think and to learn. At that moment, or what I have called 'the moment of intersubjective praxis', where because of our social acts ideas align to bring forth the new, I think that we have come into a new relationship with each other, with stronger bonds of solidarity and love being established. This lifelong process of 'what it means to experience mind', is a process of freedom and emancipation that ultimately destroys the chains of social and cultural division that have been imposed by economic and political systems.

Obviously, this type of emancipatory process is long-term and involves many events and acts that are intended to alleviate suffering and torment. At some point, a major paradigm shift or revolutionary change may occur, to rid communities of oppressive practice. But for that change to be sustained and for democratic principles of liberation and relationship to be embedded in daily life, new knowledge must be experienced and ascertained by societies as a whole. By itself, formal education will not erase social division and discrimination, but it can open up new avenues of individual and group emancipation within current worlds. It is difficult to describe those spontaneous 'feelings' of knowledge and freedom when we realise we know, when connecting with a portrait or landscape for the first time, when reading about a historical event, when someone talks about their own experience, when looking deep within another's eyes. However we know what those moments of intersubjective praxis are like when the past, present and future combine to create a new vision of what is and what might become, we suddenly see not only the objects of our thought, but ourselves differently. They are our moments of becoming free and human.

Case 3. Connecting with Tower Hill

Sandy always tried to visit her home town every few months, but it had been a couple of years now since she had managed a brief trip. She knew it was no excuse, but work seemed to be getting busier by the day and city life more chaotic. She loved the calmness of the countryside and how it brought back memories of her childhood, events that still loomed large in her mind. She never got tired of retracing her steps on each visit, whether that be a quiet coffee in the café she always went to with school friends, sitting at the netball court remembering games won and lost, or walking around the botanical gardens and talking to the trees that had always been there. Summer or winter was of little concern as she was just pleased to in familiar surroundings where she felt at home. It had been difficult for Sandy to leave her country town and go to university in the big city, but the opportunity arose and needed to be taken. She studied law and economics and got a job with a major legal firm, initially researching and providing background for senior partners. For reasons that she never really understood, Sandy had developed an interest in art from her school days and had always attended any exhibitions that were held at the small gallery in her town. She was aware of European painters who came to Australia in the early days of settlement, fascinated by the clear and bright light that they found and how they depicted the bush and landscapes with European eyes. An important painting held by the local gallery, was that of Tower Hill by the Austrian painter, Eugene von Guerard. Tower Hill was an extinct volcano a few kilometres out of town and a well-known landmark of obvious interest to von Guerard as he travelled around the district in the 1850s and 60s painting his impressions of the gold fields, rivers, waterfalls, rolling hills and the like. It was possible to walk around the craters of the volcano and feed the emus and kangaroos that lived there, something that Sandy enjoyed immensely. When standing on the rim of highest crater, you were able to get a panoramic 360-degree view of the surrounding countryside, including the nearby coastline and beyond that, the horizon, showing it seemed, the curvature of the earth. Sandy always loved that feeling, the diversity and extent of the natural environment and how small she was in comparison. What gave her a great deal of pleasure was being able to trek to the exact spot where von Guerard stood to paint his representation of Tower Hill. Over the years, the sides of the volcano had been cleared of vegetation, but in the 1970s, von Guerard's painting had been used to guide replanting. Standing on the hill and with a photograph of the painting in hand, Sandy marvelled at how what she saw before her was replicated by the painting itself. Those who had done the revegetation had to be congratulated. She wondered what had gone through the artist's mind as he had stood there as well and had observed beauty and history laid out before him, together with the warmth of the sun, the gentle breeze and the sounds of parrots and cockatoos. These were the moments when Sandy thought about her two lives, her involvement with the law and legal cases that could sometimes be emotional but often cold and remote and the enduring experiences of her childhood that brought her back to this spot year after year, almost to the point of tears. How could a painting of the past do that and an artist she had only read about? As Sandy walked back to her car, she imagined von Guerard collecting up his sketches and taking them back to his studio, perhaps thinking exactly what she was thinking now. What was it about Tower Hill that he had to capture and describe? She would have some stories to tell her friends at work next week.

4

LIVING WITH ETHICAL CONDUCT

New creative virtues are persistently being formed
by habitus in relation to the contradictions of mind
sweeping away older conceptions of being human
in struggle with new conditions that challenge love
for moral ascendency in the harsh light of day
not only during quiet moments before dawn and memory.

There is little attention given to the ethical nature of formal education itself. We could ask whether compulsory formal education through schooling is an acceptable ethical practice whereby predetermined knowledge and information is imposed by the state on children and examined with pass or fail grades? It is certainly undemocratic for any society to bring its full legal and institutional authority to bear on what children should think. My attention to such matters was first alerted when I began teaching and became a cog in the curriculum and assessment machine. I was not entirely clear on what I thought about schooling generally, but I believed that imposing viewpoints on children was not correct. I quickly became involved with the introduction of desk-top computers into schools and my reading of Papert (1980, 1996) in particular showed me that we were dealing with philosophical rather than technical issues. Early books by Hubert Dreyfus (1972/1992) and by Joseph Weizenbaum (1976/1984) regarding the emerging field of artificial intelligence and their views on the problems of seeing connections between what Weizenbaum called 'computer power and human reason', made a big impression on me. As a teacher, I wanted the new computer-based technologies to expand the knowledge horizon for children, but within an appropriate ethical framework. Weizenbaum wrote that the assumptions being made by artificial intelligence about how humans think could be giving an entirely inaccurate view of our humanity, with the brain being seen as an 'information processing machine' a case in point. He anticipated the time when computing and technology generally could make possible human activity that is morally reprehensible, such as head transplants and that if so, should be pre-empted now. If artificial intelligence could result in outcomes that we consider repugnant and disgusting, then that field of human endeavour should be stopped. For these reasons, I wrote some brief articles that referred to what I called 'technoethics', pointing out that artificial intelligence might be the first time that human exploration was discontinued because of philosophical objections. Since then of course, computing and artificial intelligence have become central components of military and industrial applications with little regard for ethical guidance. An unfortunate current example of how technology connects the social and military/industrial is that of drones, where their application in war and surveillance today and current discussion of autonomous 'killer drones' is disturbing but common (Hastings, 2012). I am not aware of similar debate regarding social media and whether its use should not be allowed in schools for similar philosophical reasons.

Over more recent times, my thinking on ethics has been stimulated by my reading of Aristotle and of Levinas (Strhan, 2016). I shall comment briefly on Aristotle later and his significance in relation to practice and praxis. My attention to Levinas was drawn by Biesta and his discussion regarding the 'self' and education. Biesta (2017a, p. 43) argues that traditional education sees the child as 'an object of the teacher's intervention, but never as a subject in its own right' and that 'our subject-ness is not constituted through acts of signification'. Similarly, Levinas approaches human subjectivity from an ethical standpoint and the view that 'responsibility' is 'the essential, primary and fundamental structure of subjectivity' (Levinas, 1985, p. 95, cited in Biesta, 2017a, p. 44). From this, I take it that an ethical state of responsibility, or responding to the other, exists before acts of meaning and knowledge occur, existing as a natural mode of subjectivity. In Chapter 2, I nominated a set of ten characteristics of being human, or human being, including subjective and knowledgeable, from which human existence is constructed. I find it difficult to separate these ten characteristics and prefer to interpret our humanness as an integrated whole that exists in and with the world of experience. In addition, use of the term 'signification' above reminds me of the work of Peirce on semiotics. Peirce (2015b, p. 99) stated that 'A sign, or representamen, is something which stands to somebody for something in some respect or capacity. It addresses somebody, that is creates in the mind of that person an equivalent sign, or perhaps a more developed sign.' From this, he developed three notions of signs or phenomena as Firstness, or qualities of the sign itself, Secondness, or characters that the sign exhibits and Thirdness, or depending on how it is interpreted, expressing possibility, fact or reason. A speech utterance for example has qualities of tone and strength, may suggest anger or sadness and can indicate what is expected. Looked at in this way, there does not seem to be any major contradiction between Levinas and Peirce, except in the origin of the 'responsibility' impulse. That is, while subjectivity is part of our wholeness, it does not necessarily 'come first' in how we engage the world, not requiring knowledge of previous understanding. I would submit, tentatively, that characteristics of ethics and responsibility are not fundamental or original but are socially derived from political, cultural and personal experience. This does not resolve the question of human objectification compared with subjectification, but it does locate it within the realm of sociality, impacted by human reflective action and discourse. It also connects issues of ethics, responsiveness and love to how we conduct our own lives, a dialectic between inner and outer, rather than being dominated by foreign control.

I have received some criticism that my approach to education relies on a male experience – Kant, Hegel, Marx, Dewey for example – an issue that will be taken up in more detail in Chapter 10. However it has been my experience and narrative for which I do not have a need to apologise. My reading of Dewey could be seen as the centre-point of my personal educational trajectory, extending to American Pragmatism, Mead, Freire and Vygotsky as key nodes of understanding. When connections were made between practice and praxis for my own teaching, a new world was opened up regarding Greek philosophy and in particular, Aristotle. I found Aristotle's division of knowledge into the productive, practical and contemplative useful when thinking about how to involve students with mathematical and scientific knowledge and indeed, how little discussion there was among the profession of such matters. As mentioned previously, the concept of happiness or well-being intrigued me, although I was far too busy for it to be all consuming. I noted that Aristotle thought happiness could not be evaluated over a short time and that it required contemplation within a specific scaffold. In relation to happiness, I discovered that he had written about ethics and had described a series of excellences or virtues by which we could reflect on our lives. This concept of 'virtue ethics' appealed to me, although I found it very frustrating in trying to relate them to my own life to date; it was almost impossible to consider these virtues in response to a set of criteria, or in neoliberal language, key performance indicators, that could be clearly supported with evidence.

Aristotle categorised virtue as moral or intellectual with the former including courage, temperance, justice and the like and, to live well, an understanding of how these approaches fit together cohesively. He identified 'practical wisdom' as the way we do this, not through taking up general rules as such, but by proceeding with social practice so that our understandings of well-being can be enacted appropriately. There was a 'golden mean' of conduct that existed with each virtue, so that we did not drift between excess and deficiency. For example, with courage, we need to act between foolhardiness and cowardness. My teaching was influenced by these ideas, fitting nicely with pragmatism, where I was able to encourage my students in mathematics and science to act with courage in pursuing their ideas, to make judgements about what they were observing and how they might proceed, to be generous in sharing their thoughts with others, to be modest and friendly in their approach. Aristotle's 'virtue ethics' then suited my understanding of knowledge production, rather than other views that saw the ends justified by the means (consequentialism, judged by outcomes), the means justified by the ends (deontological, or duty and rules) and the greatest good for the greatest number (utilitarianism, utility minus suffering). These were important ideas, but the notion of 'virtue' being in relationship with 'prudent practical wisdom' was a practice I could work with every day in class. Again, I have taken a lead from Biesta (2013, p. 137) in extending the idea of virtue to that of 'educational virtuosity', described not as 'a set of skills or competences but rather as a process that will help teachers to become educationally wise'. I have attempted to identify a set of six educational virtues of practice myself (Hooley, 2018, p. 164) comprising dialogue, inquiry, respect, judgement, responsibility and reflection. In my view, these connect with Aristotle, Biesta and perhaps even Levinas in describing what could be called 'classroom subjectivity', that is subjectivity contextual for the culture and knowledge of formal schooling. I should point out that 'classroom subjectivity' is not an abstract concept that has a pre-existence, but is organic, constructed by teachers and students working together in their efforts to understand the world of education. It is constructed through social action over periods of time as the meaning of schooling, curriculum, authority and knowledge are experienced (Hostetler, 2016). It does not fall from the sky, unannounced, but is built cognitive structure by structure using collective consciousness, language, history and community aspiration in relation to what we take to be real and meaningful.

Signature pedagogies and ethical life

We now turn to considering how the notion of 'educational virtuosity' and 'classroom subjectivity' can be attempted in practice. Since the mid-2000s, our research has identified eight signature pedagogies of praxis teacher education (Table 4.1) around which programs can be designed and implemented and by which preservice teachers can investigate the realities of teaching and learning (Arnold *et al.*, 2012; Hooley, 2018). Given that the signature pedagogies are based on practice, partnership and praxis they, in total, constitute a framework of ethical conduct. As noted above, this approach draws explicitly on Greek philosophy and the concept of virtue ethics, concerned with what is right and to live according to what is appropriate for specific circumstances. Practical or prudent reasoning enables character to be built from personal action and the judgement to act in accord with what is considered to be right for teacher and student participants. Over long periods of time, the result of human action is agreement around a set of excellences or virtues that enable a life well lived particularly for the public good.

Education and teacher education conducted in this way, around praxis and autonomous action, allows meaning to emerge from social acts and reflection on outcomes. This process is socially and educationally equitable because it respects and values background culture and understandings of all participants and considers all views as they are constructed as valid for those involved. Community reflection and critique on viewpoints enables more generalised thinking to be appreciated, not only in regards other immediate participants but in relation to past experience

TABLE 4.1 Signature pedagogies of praxis teacher education

Signature pedagogies	*Characteristics of signature pedagogies*		
Professional Practice (Schatzki, Kemmis, Green)	Recognises personal learning from immersion in practice	Supports communities of practice to support inquiry for improved new practice learning environments and student learning	Continuing critique of practice for change of conditions to formulate ideas of
Repertoires of practice (Kalantzis, Cope)	Identifies and articulates features of pedagogical, curriculum, assessment practices	Links key features of pedagogy, curriculum, assessment for change and improvement	Critiques repertories of educational practice as social activity that supports satisfaction and progress
Teacher as Researcher (Stenhouse)	Systematically investigates own practice for improvement	Participates as member of school-based research team/s	Relates local, national and global research, policy and practice
Case Conferencing (Shulman)	Generates case and commentary writing for understanding of practice	Participates in case conferencing and concept analysis for production of teachers' knowledge	Encourages articulation and analysis of teachers' knowledge in relation to theories of curriculum and teaching
Community Partnership (Gonzalez, Moll and Amanti; Sizer)	Connects with local communities	Integrates community culture and knowledge into curriculum	Investigates community to understand local aspiration, history, knowledge, language
Praxis Learning (Freire)	Investigates/ provides description, explanation, theorising and change of practice in response to reflection on practice	Demonstrates a curriculum developed from praxis and in response to reflection	Constructs learning environments of ethically-informed action for the public good
Participatory Action Research (Kemmis, Brennan)	Identifies and advocates key issues of policy and participates in collecting data for analysis	Contributes to project discourses with internal and external team members	Theorises and critiques research findings in the public domain
Portfolio Dialogue (Freire, Dewey, Brookfield)	Compiles and discusses artefacts of personal learning over time	Participates with artefact and knowledge discourses that show understandings of meanings of practice	Demonstrates a coherent philosophy consistent with personalised practice and community change for public good

and the literature as well. Praxis teacher education arranged around the signature pedagogies is therefore, of itself, ethical and equitable in intent and immerses all participants in processes of this type, rather than teaching about them somewhat remotely and abstractly. Discussions with preservice teachers and with university staff indicate preliminary awareness that the signature pedagogies assemble a framework of 'practical wisdom and judgement' such that all participants are placed in the position of acting for the social and educational good. This is fundamentally and philosophically different to merely being expected to passively reproduce existing knowledge and practices.

Each of the eight signature pedagogies provides preservice teachers with a unique lens in which to explore ethical standpoints connected to educational contexts. Case writing and conferencing encourages preservice teachers to select and describe a key incident that involves a dilemma, complication or accomplishment (Shulman, 1991). Through this frame preservice teachers are able to reflect on teaching and learning practices that display a social and moral commitment to learners and their families. Questions can also be formulated around ethical standards for the teaching profession, ethical professional relationships and ethical institutional footprints.

Deliberation on the consequences of unethical actions can also prompt preservice teachers to wonder about:

What could be otherwise here?
Whose lifeworld is being represented?
What are the political, social, cultural and institutional forces at play?
What structures enable or disable human agency?

Teacher as researcher, the third signature pedagogy of praxis teacher education, asks that preservice teachers engage in a range of research practices that facilitate deep systematic inquiry; inquiry that is made public and shared with others (Stenhouse, 1983). In this space preservice teachers interrogate and investigate research questions that are situated in school-based experiences. Global themes, curricular resources, pedagogical approaches and strategic policy directives can be explored and critiqued through the collection of research data. In the generation of data findings and through engaging in collaborative dialogue, broader perspectives can be gained on ethical standards that underpin teacher professionalism. Additionally, generalisations, ethnocentric viewpoints, patriarchal politics and specific attitudes of race and gender that impact on successful learning can be exposed.

Teacher morality and ethical decision making is more likely to surface when preservice teachers grapple with those schooling elements that seek to undermine, exclude, oppress and marginalise even further those individuals who are the least advantaged (Connell, 1993). Portfolio Dialogue, the eighth signature pedagogy, links preservice teachers' skills and knowledge with new learning journeys. The inclusion of school-centred artefacts, annotations and philosophy statements in the portfolios triggers collaborative conversations around questions like, 'What does it feel like to be a learner in my classroom?' 'How do I do what I do better?' In the case of preservice teachers, formulating an ethical stance is more likely to develop when oppressive forces surrounding teaching experiences are dismantled and found out. Participating in professional dialogue with peers, teachers, community stakeholders and lecturers can also help to reframe taken for granted assumptions of one's practice. Darling-Hammond and Richardson (2009, p. 50) acknowledge that change can only emerge when practitioners open up and expose themselves to opportunities for critique in the work that they do, when:

> . . . group members must make their practice public to colleagues and take an inquiry stance. Change occurs as teachers learn to describe, discuss, and adjust their practices.

Acquiring an ethical practitioner stance therefore requires a willingness on the part of preservice teachers to expose their vulnerabilities and uncertainties. When this is done in a safe way through portfolio dialogue, a more sophisticated understanding of ethics and equity can ensue.

For signature pedagogies to map the key features of lifeworld for investigation and change, as indicated by discussion of ethical conduct above, they must also be able to deal with another major feature of society and education, the competing issues of power and equity. That is, signature pedagogies must be able to establish a learning environment that is equitable for all participants and is not dominated by discrimination, prejudice, bias and coercion. This means working within a context of democratic respect, autonomy and cultural expression. Under capitalism, this is extremely difficult to achieve, as power is a defining feature of economic systems that maintains private privilege rather than public benefit at both individual and institutional levels. Power seeks to exclude rather than include all who are involved. As a starting point for enabling equity and taken as a whole, the signature pedagogies encourage a broad experience of society and of learning such that participants do not remain comfortable within their current sphere of understandings but must reach out to the unfamiliar and unclear and draw upon their intuitive knowledge to explain the new. This requires being able to read well the new field of experience and not be too alarmed or discouraged at what is found.

Confronting emergent experience means deciding on fresh and innovative social acts that provide avenues into new options of thought. Educational research for example can adopt non-traditional and creative methodologies that bring into focus a wide variety of data sets that not only scaffold investigation of particular questions, but also enable critical reflexive considerations by researchers of their own intersubjective understandings of research and knowledge, a process of intersubjective praxis. For example, a colleague related the story of a school student researching about medieval concepts found a family connection to her great grandfather who lived in the ninth century in Spain. The rose, which appeared on all 167 books her scholar great grandfather wrote, was at the centre of her model and presentation to manifest her understanding of the Jewish Golden Age in Spain at the time. Conceptualising and problematising research as inherently creative and critical in this way supports ethical and equitable conduct. Currently, the map of signature pedagogies does not adequately or explicitly deal with the issue of power and equity and additional items are required. A draft set of possible items is shown in Table 4.2, but these are still to be validated at this time and are presented for discussion only.

We theorise that as well as providing a map of practices regarding learning, teaching and the organisation of knowledges, the signature pedagogies of praxis teacher education enable ethical and equitable conduct for all participants. They establish the conditions for general and educational virtues to enhance learning, intellectual and social virtues such as kindliness, empathy, concern, compassion, fairness and the characteristics of democratic inquiry (Hopkins, 2014). In this way, the signature pedagogies encourage ethical and equitable acts through personal and group action and interaction, rather than being merely taught about abstractly. That is, the map of signature pedagogies describes a lifeworld that can exist and be created in classrooms and the broader society as well.

There are deeply philosophical questions here that are frequently overlooked in educational debates. For example, we need to be able to defend the issues of ethics and equity themselves as issues central to educational purpose, but to do so not in terms of systemic imposition on young minds, but in relation to the place of education for all people in becoming more human, more just and more virtuous. Harrison and Bawden (2016) in adopting a generally Aristotelian position, describe virtues as 'settled (stable and consistent) traits of character, concerned with morally praiseworthy conduct in specific (significant and distinguishable) spheres of human life'. On this basis and according to the philosophy of phronesis and praxis, virtuous conduct emerges from the many relationships that humans have with their environments and how new events and understandings are incorporated into ongoing experience. Designated as the 'principle of sociality' by Mead and indicated below (Mead, 1908), this underscores the pragmatic view of what it means to be human, as all of us relate to each other in the plurality of social experience and come to appreciate the nature of existence within the context of historical and temporal socio-cultural being:

> Now, to a certain extent the conception of an evolution of environment as well as of the form has domesticated itself within our biological science. It has become evident that an environment can exist for a form only in so far as the environment answers to the susceptibilities of the organism; that the organism determines thus its own environment; that the effect of every adaptation is a new environment which must change with that which responds to it. The full recognition, however, that form and environment must be phases that answer to each other, character for character, appears in ethical theory.

TABLE 4.2 Signature pedagogies of praxis teacher education with equity column

Signature pedagogies	*Characteristics of signature pedagogies*			
Professional Practice (Schatzki, Kemmis, Green)	Recognises personal learning from immersion in practice	Positions participant interest as central concern without bias	Supports communities of practice to support inquiry for improved learning environments and student learning	Continuing critique of practice for change of conditions to formulate ideas of new practice
Repertoires of practice (Kalantzis, Cope)	Identifies and articulates features of pedagogical, curriculum, assessment practices	Adopts mix of innovative practices to meet specific needs	Links key features of pedagogy, curriculum, assessment for change and improvement	Critiques repertories of educational practice as social activity, progress and satisfaction
Teacher as Researcher (Stenhouse)	Systematically investigates own practice for improvement	Recognises research as situated in participant experience	Participates as member of school-based research team/s	Relates local, national and global research, policy and practice
Case Conferencing (Shulman)	Generates case and commentary writing for understanding of practice	Authorises narration and commentary of lifeworld case and story	Participates in case conferencing and concept analysis for production of teachers' knowledge	Encourages articulation and analysis of teachers' knowledge in relation to theories of curriculum and teaching
Community Partnership (Gonzalez, Moll and Amanti; Sizer)	Connects with local communities	Ensures partner relationships are 'without prejudice'	Integrates community culture and knowledge into curriculum	Investigates community to understand local aspiration, history, knowledge, language

continued . . .

TABLE 4.2 Continued

Signature pedagogies	*Characteristics of signature pedagogies*			
Praxis Learning (Freire)	Investigates and provides description, explanation, theorising and change of practice in response to reflection on practice	Supports autonomous, non-coercive practices	Demonstrates a curriculum developed from praxis and in response to reflection	Constructs learning environments of ethically-informed action for the public good
Participatory Action Research (Kemmis, Brennan)	Identifies and advocates key issues of policy and participates in collecting data for analysis	Encourages participation of all cultural backgrounds	Contributes to project discourses with internal and external team members	Theorises and critiques research findings in the public domain
Portfolio Dialogue (Freire, Dewey, Brookfield)	Compiles and discusses artefacts of personal learning over time	Assists new praxis through problematising experience, themes and actions	Participates with artefact and knowledge discourses that show understandings of meanings of practice	Demonstrates a coherent philosophy consistent with personalised practice and community change for public good

In this passage, Mead summaries that ethical conduct is constructed as we speak to the world and the world speaks to us.

Education as ethical intersubjective praxis

Living an ethical life, I suggest, is first of all an 'action' structured by our humanness and guided by our political and cultural circumstances and second, proceeds 'socially' as we conduct human being and engagement with the world. Bearing on this view and as has been pointed out by Hansen (2006, p. 165), Dewey's final sentence in his epic *Democracy and Education* integrates a number of his theoretical ideas: 'Interest in learning from all the contacts of life is the essential moral interest' (Dewey, 1916, p. 360). I argue that 'interest in learning' stems from my ten characteristics of being human above and involve establishing relationships with the environment to resolve the situation at hand. For this resolution to be successful, human being must participate with 'all the contacts of life' whatever they may be, regardless of their challenge and difficulty. It necessarily follows that all contacts arise from community experience and communication, rather than individual and private practice. Finally, this process of continuous social act produces a mode of existence that is appropriate and beneficial for the organism. Because the process is communal, contacts will be both negative and positive, but in totality and ultimately, will be in the interests of the species as a whole. I want to use the term 'knowledge' at this point, to suggest that it not only denotes a body of understanding but also is the linking process by which our rendezvous with the world occurs. We could instead consider 'teaching and learning' in this way, but regardless of how they take place, it seems that their product is what counts. That is, we may have excellent formal teaching about the principles of Darwin's evolution, but it is what we take with us as perspectives, opinions, confusions after leaving the classroom that structure our relationship and exchanges with the social conditions we confront.

Acting ethically in a supportive, neighbourly and responsive manner such that the human interests of community are enhanced, endures through knowledge production, in accord with Gidden's view of 'all humans are knowledgeable agents' (op. cit.). Peirce is said to have adopted the notion of 'community of inquirers' to describe the location of this process, by which he meant that 'all people belong to a larger-than-national community of thought' (Burgh, Field and Freakley, 2006, p. 32). This is a further pivotal stance of pragmatism that shows individual experience and thought existing within a much larger frame of historical and cultural life, not isolated and relative, but immersed and public. Communal values and beliefs emerge in this way. Similarly, we can consider the powerful human emotion of love as a paradigm and pragmatic expression of ethical knowledge and experience that has been constructed over many centuries from 'all the contacts of life'. The Freirean scholar, Darder (2015, p. 47) for example, opens up discussion on love by the following quote from Freire (1993):

> I think that it could be said when I am no longer in this world: 'Paulo Freire was a man who lived. He could not understand life and human existence without love and without the search for knowledge.'

It is unusual for an educator to speak of love in this way, although in terms of schooling, I think that Freire had a broader than usual vision. His strong sense of unity with the oppressed and the dignity of all people regardless of background, establishes a sense of solidarity between parents, teachers and students for liberation. Fromm's construct of love discussed above involving care, responsibility, respect and knowledge can be seen in all democratic classrooms especially I would argue in how knowledge is anticipated. It is difficult to envisage dignity and equity when knowledge is specified and imposed from without and there is little opportunity for experimentation and practice. According to Darder, Freire believed that as we live, learn and love together, we come to understand that knowledge is co-created and that therefore we are historical beings each with our role to play in forging emancipation. There is a challenge here for teachers and educators, given it is commonly said that we can be friendly with students, but not friends. This is particularly so in many countries now where harassment of various types has become a litigious issue with considerable publicity. Teachers are strongly advised never to touch children even when they might hurt themselves in the playground or become upset for some reason. If the law of the land is such that teachers must restrict their involvement with students then that must occur, but that does not prevent the conditions for solidarity and friendship being established from a knowledge perspective. I think the point being made here is that parents, teachers and students can become authentically known to each other as best they can within the constraints of curriculum, assessment and institutionalised authority. Social class friendship can be assembled. This can be an ethical relationship, where all participants are engaged in the active process of knowledge, of interacting with the world experientially in all its aspects, of listening to what the world has to say and of changing ourselves accordingly.

Our discussion so far in the first four chapters of this book has emphasised philosophical issues regarding human existence including dialectical materialism, knowledge generally and in the sciences and humanities specifically, practice and praxis, the notions of subjectivity and intersubjectivity and the nature of ethical conduct. We shall work with these ideas throughout remaining chapters. Constituting a philosophy of practice, these features all impinge on Mead's 'social conception of education' and are in marked contrast to the underpinnings of neoliberal schooling. Let me add some brief explanatory detail at this stage to three other concepts that I have mentioned and their positioning in schooling of a new type, those of educational virtuosity, classroom subjectivity and intersubjective praxis. I have drawn upon Biesta's view of educational virtuosity because of its links with the virtue ethics of Aristotle and how I envisage the dialectic between action and thought. Defining educational virtue as arising from excellent social practice, or 'excellences' as described by Aristotle, that is practice that seeks to enable the liberation of others, we can argue for a vision of schooling that is human-oriented rather than self-oriented. My suggestions above for educational virtue such as dialogue, inquiry, respect, judgement, responsibility and reflection, can all realistically be implemented at all levels of education and help restore the dignity of knowing (Taylor, 2016). My next step was to extend this understanding of educational virtue to what I have called 'classroom subjectivity', a term that is intended to convey the meaning of the organisation within schools as a collective understanding of what classroom are all about: groups of teachers and students working together on significant issues to generate tentative knowledge for ongoing experiment. In this way, there develops an awareness or consciousness, that knowledge is socially constructed from social groups of interest and that this subjective awareness of classroom life can be generalised and incorporated elsewhere.

There is little doubt that 'classroom subjectivity' already exists, but it is neoliberal, where all participants are aware that they are pawns in the game of passively accepting slices of information that have been imposed on them. Formal systems of examination rigidly support this view. To counter such rigidity however the notion of 'to knowledge' as intersubjective praxis can be established in practice. Intersubjective moments occur from praxis when there is close alignment of experiences from past and present encounters to bring forth the new, novel understandings of events and situations that were not held before. In some cases, the new alignment may seem unrelated and can be called creative, whereas on other occasions, the new gives an insightful explanation and can also be called creative. We can observe this in classrooms all the time, when students struggle to comprehend what is presented and come up with their own hesitant conclusions. Our discussion thus far is also grappling with speculative possibility and conclusion that may crystallise from the main ideas we have considered, but the general direction is clear. In all societies, formal education is an important part of political and economic ideology and adheres to a distinct moral and ethical system. There is a dialectical relationship between moral and ethical constructs, between overall agreement on what is right and wrong and what actions need to be taken in particular situations to live in accord with each other (Keane, 2016). These questions are influenced by the historical and socio-cultural circumstances of communities and need to be refined over time, through social practice. I suggest that it is the knowledge question, or the process 'to knowledge' and of becoming human in thought, deed and awareness that lies at the heart of this philosophical dilemma.

Acknowledgement. Discussion of signature pedagogies for praxis teacher education in this chapter draws on the work of the Praxis Research Group over a number of years within the College of Arts and Education, Victoria University Melbourne.

Case 4. To chat or not to chat

'Can I go please miss, we are playing finals this week?' Wednesday mornings were often a little disruptive as it was sports' day and every so often, teams were getting organised for inter-school games in the afternoon. Nicole had only been teaching for a few years and was still finding her feet, but she was feeling more comfortable at walking into most of her mathematics and science classes at any level. She had a job in a large multicultural secondary school in the inner city and the complications of working with a diverse staff (and an idiosyncratic principal) of nearly seventy had its moments. However she found the students likeable and interesting as they generally took her advice about the topics that had to be covered and in preparing for the inevitable unit tests that were due every 6 weeks or thereabouts. It was difficult she thought to always follow the curriculum when her students might have other interests at that time. Nicole usually looked forward to this particular group of Year 9 students on Wednesdays as they were pleasant enough and worked fairly well in the small groups that she arranged. She always had her sessions well planned and tried to include some type of activity, even if only for a few minutes, to get away from the dominance of the text book. If fact, she found the recommended texts in mathematics uninspiring and was constantly searching through books and online to compile her own units consisting of interesting situations and problems to which students could relate as closely as possible. On this particular morning, Nicole smiled to herself as the group came in chatting among themselves as usual. They had plenty of energy and it was common for her colleagues to tell her that Year 9 was often difficult, a transition year, as students changed from being the youngest in the school, to thinking about their futures and subject selection for the senior years. She liked this age group of mid-teens and could joke with them or be very serious as the work demanded. What worried her most was that they seemed to take a long time to settle down each class, generally talking about what was happening at school or elsewhere, problems with their friends or family, where they might be going on the weekend, how their sporting teams were going and the like. This seemed a waste of time to Nicole, especially when tests or exams were on the horizon. Whatever she tried it didn't work. She was very reluctant to impose new rules and anyhow, rules that were imposed often had the opposite effect to that intended. After a few minutes, she tuned in to the conversations and realised that interspersed were comments about the worksheets that she had distributed on the tables: 'What's this about?' 'Does this follow on from last session, where are we up to?' 'I'll get the equipment from the cupboard?' 'I saw something about this on the Internet the other day?' It suddenly occurred to Nicole that perhaps their chattiness was not time wasting at all, but a way of getting organised and thinking about the task she had set. She didn't see the connection between that comment about Billy's big sister and triangles, but there may be one that would become clear later. Perhaps they had all these ideas whizzing around in their heads that didn't make a lot of sense just now, but it was important that they explore them and be able to work with them. Nicole's head was whizzing with new thoughts as well and she had to remind herself that all ideas and their expression needed to be respected in school if she was to support all of her students delve deep into their own experience. She resolved to go with the flow.

Knowledge Exemplar 1

At the conclusion of Parts I, II and III of this book, I will include a 'knowledge exemplar' as a means of summarising the key ideas that have been encountered. In this, I am drawing on the work of Kuhn (1977, p.187), where in discussing knowledge and what he termed the 'disciplinary matrix' of science, he writes that, 'More than other sorts of components of the disciplinary matrix, differences between sets of exemplars provide the community fine-structure of science' (cited in Hooley, 2010, p. 189). A more expansive discussion of matrix exemplars can be found in Hooley (2010, pp. 188–212). I have therefore briefly outlined Knowledge Exemplar 1 in Table 4.3 as a matrix of key features of formal schooling and a series of indicators that would demonstrate their presence in primary or secondary schools, or indeed university courses and family activities. In considering the exemplar as a whole there is a very clear picture of schooling as 'philosophy of practice', where the process 'to knowledge' involves human action of engaging, reflecting and interpreting the world. This is a dialectical social process of collaborative experience, where all of us of different cultures, ages and positions relate and interrelate with the world in a human way as we construct and reconstruct what it means to be human.

TABLE 4.3 Knowledge Exemplar 1: formal schooling

Social acts	*Indicator 1* *Curriculum*	*Indicator 2* *Pedagogy*	*Indicator 3* *Assessment*	*Indicator 4* *Research*
Dialectical Materialism	Interconnections always made between topics, ideas	Reflection on experiments, activities for possibility	Discussion of outcomes for connections, options	Proposals for new studies of change, transformations
Practice, Praxis	Knowing arising from doing for public good	Learning circles for collective wisdom on projects	Learning circle consideration of project artefacts	Sharing of outcomes for ongoing investigation
Intersubjectivity	Social awareness, understanding of knowledge	Discursive, language-based communication	Negotiation of group outcomes for consensus	Identification of collective interest for future projects
Virtuosity	Recognition of features of humanness	Agreement on key virtues as basis of approach to teaching	Informal awareness of educational virtues	Compare, contrast different features of formal schooling
Ethical Conduct	Engage education for public good	Democratic in interests of all participants	Non-coercive, respectful of all views	Discussion of different ethical understandings
To Knowledge	Participation with practice-theorising	Holistic, enabling culture, history, language experience	Narrative accounts of projects and processes undertaken	Investigate knowledge as process, not knowledge as given

PART II

Looking inwards, looking outwards

Modern science is still grappling with how the universe functions and utilises a combination of explanatory models that are based on experiment and reason. It is possible to describe the orbits of planets in terms of force and gravitation and the composition of substances in terms of particles or electromagnetic waves of various types. Generally speaking, philosophy also operates with broadly descriptive models of the mind, although much detail remains to be agreed. Taking some of the ideas encountered in earlier chapters therefore and seeking to understand human learning at deeper levels beyond the general model can be a frustrating process when 'thinking about thinking' reaches an intellectual dead-end. Part II of *Dialectics of Knowing in Education* however focuses on the question of epistemology and how we currently understand the human mind connects with social and physical environments and makes sense of what is considered to be real. This generates many philosophical and metaphysical problems regarding knowledge, thinking and thought, human consciousness, emotion and the nature of language. A discussion of human subjectivity illustrates the difficulty of definition as the human subject attempts to understand itself. Consciousness is a similar concept and occurs as a result of being human, in the same way as eyes, legs and fingers appear and grow. Language and creativity can be considered in the same category as characteristics of being human that enable tentative connection with the social and physical worlds. Linguists such as Wittgenstein, Chomsky and Pinker offer insights as to the question of mind and therefore how we learn and comprehend remains unresolved. Dewey's notion of technology as 'productive act' in generating thoughts, artefacts and further acts is also discussed as a means of theorising how all participants experiment with what they know and how they might know something else. It is argued that educators at all levels need to take into account such issues so that they have a generalised framework of acting when they walk into every class and attempt to engage students. That is, 'looking inwards, looking outwards' enables a flexible understanding of each person and of the group as they attempt to consider what the teacher, curriculum, or friend has specified and bring their personal understandings to bear on new situations: 'What should we do now?' While the philosophical questions of subjectivity, consciousness, creativity and language remain mysterious, they are at the centre of community, family and personal life and therefore at the centre of what it means to be knowledgeable citizen, comrade and human.

Note. Part II includes a selection of schema as genuine and informal electronic posts to indicate my public thinking at particular times in relation to chapter issues. These are included as authentic descriptive and narrative data of 'looking inwards, looking outwards' as I read, write and grapple with the difficult conceptual problems involved in preparing this book.

5

SUBJECTIFICATION

How we are

> New creative virtues are persistently being formed
> by habitus in relation to the contradictions of mind
> sweeping away older conceptions of being human
> in struggle with new conditions that challenge love
> for moral ascendency in the harsh light of day
> . not only during quiet moments before dawn and memory.

A central economic imperative of capitalism and neoliberalism is the abstraction of major characteristics of humanness, followed by their quantification and finally their allocation of certain value to the production line. In this way, the human workforce is 'objectified' as mere objects of production, devoid of all aspects of compassionate humanity. According to Lukacs (1922/1968), this process of reification leads to alienation not only from the outputs of production, but from what it means to be human itself. Formal education has a key role in reversing this process, to counter dehumanising objectification and to replace it with 'subjectification' (Biesta, 2013, pp. 84–85). This latter process will need to emphasise:

- respect for the autonomous action of all persons;
- the culture, character and knowledgeability of people from all backgrounds;
- the need for discursive community environments for the development of views and practices and the establishment; and
- consolidation of democratic public spheres for mature and non-coercive free assemblies and debate.

Broad subjectification principles of this type then need to be adjusted as need be for application in educational institutions and public schools at all levels, such that the main purpose of schooling becomes the subjectification impulse. This is a difficult task under current socio-economic conditions of course (Ball and Olmedo, 2013). One way of thinking about this process is to concentrate on the human subject in terms of 'Who am I?' although this can overly psychologise the subject and bring identity politics (Fraser, 2009) into play. Post-structuralism in particular has influenced the ascendency of feminism and ethnicities in comparison to social class as the major sociological focus of analysis. In contrast, the concept of 'How' rather than 'Who' proposed by Biesta (2013, p. 18), expressed as 'How I am' or 'How we are' links strongly to pragmatism and practice and the ways in which social action is taken to construct the selfhood and lifeworld. The formation of subjectification therefore builds on the notion of 'I think, therefore I am' (action), rather than 'I am, therefore I think' (existence).

In discussing Knowledge Exemplar 1 (Chapter 4), I wrote that the social act 'to knowledge' could be understood as involving

> human action of engaging, reflecting and interpreting the world. This is a dialectical social process of collaborative experience, where all of us of different cultures, ages and positions relate and interrelate with the world in a human way as we construct and reconstruct what it means to be human.

These are difficult concepts and we use words to describe or model them at the same level as the concepts themselves. For example, descriptive words are used regarding my concept of 'to knowledge', explaining it as a process of social action interacting with a situation we encounter so that the situation can be changed in such a way that further action by us is possible. We can think of adding some salt to a container of soup to change the taste more to our satisfaction, an act that changes the soup and changes our relation to the soup at the same time. If the Indicators used to describe various aspects of 'to knowledge' are considered (Table 5.1), it is possible to come to a deeper appreciation of the process being suggested and perhaps more importantly, the overall, integrated outcome as far as the human organism is concerned.

TABLE 5.1 Indicators of social act 'to knowledge'

Social Acts	*Indicator 1 Curriculum*	*Indicator 2 Pedagogy*	*Indicator 3 Assessment*	*Indicator 4 Research*
To Knowledge	Participation with practice-theorising	Holistic, enabling culture, history, language experience	Narrative accounts of projects and processes undertaken	Investigate knowledge as process, not knowledge as given

Let us ponder the soup example above. Each person will approach the taste problem from the point of view of their own perspective, whether they have extensive experience of soup or not and will be aware that they are required to make a judgement about quality. They have an issue to resolve. Each person will respond on the basis of flavour, consistency, temperature and the extent to which certain components of the soup are discernible and acceptable. Based on this combined experience, some suggestions will be made about changes to be made, some carrots and onion should be added, a few slices of tomato are required, a sprinkle of salt and pepper. Various comments will be made with other family members offering advice from their different palates and personal perspectives. Interestingly, each indicator in Table 5.1 has been enacted. In the end, it has not been the concept of 'soup' that was important, but the experiential process by which the soup was experienced and changed. No matter what the experience of each person prior to this event, they each come away with a different understanding of the world of food and their place within its boundaries. We could say that the process 'to knowledge' has created a new individual subjectivity, a new intersubjectivity of all participants in the kitchen and a new consciousness. This is our 'human way' of being in the world.

Mead puts forward a similar argument when he writes about 'knowing' as an undertaking not always occurring within situations of 'ignorance or uncertainty' which 'knowledge' is required to 'dissipate'. He goes on to comment (Mead, 1932/2002, p. 91):

> Knowledge is not then to be identified with the presence of content in experience. There is no conscious attitude that is as such cognitive. Knowledge is a process in conduct that so organises the field of action that delayed and inhibited responses may take place.

With this insight, Mead is distinguishing between 'knowing' and what we generally take as 'knowledge' of something, or knowledge content such as Paris, elephant, mother, soup. He introduces the notion that 'conscious attitude' is not cognitive, by which I presume he means that our attitude towards Paris, elephant, mother or soup does not occur through reason or thoughtfulness, but through some other process. Mead denotes this other pathway as knowledge, 'a process in conduct', that enables the situation we are encountering to be altered such that we can respond to changing circumstances. By so doing, Mead has reconstructed the nature of epistemology itself regarding the nature of knowledge, where knowledge comes from and how it is possible for humans to engage the world in a knowledge or human way. I want to propose therefore that the social act 'to knowledge' is the basis of our subjectivity and intersubjectivity, meaning that we need to think about what we have in our heads as subject content, not as knowledge, but as information and that the process 'to knowledge' involves working with this information to create new perspectives of ourselves and of the world. As mentioned above, these are all words at the same level of complexity, but they begin to distinguish between the various forms of social acts that we undertake and their outcomes. In general, then, I am comfortable with the notion of 'to knowledge' as an act in its own right, arising from social acts of individuals and communities whereby all humans regardless of age or socio-cultural background continuously construct and reconstruct an awareness of themselves and of others, to guide further action. Figure 5.1 suggests this relationship.

I have brought together in Knowledge Exemplar 1, a matrix depiction of ontological and epistemological acts in a dialectical relationship. Figure 5.1 now portrays the result of this dialectic as 'to knowledge', our human engagement with the world. Exactly how the process of 'to knowledge' occurs in the brain must remain speculation at this point, awaiting progress in philosophy and neuroscience. Dewey, Mead and other pragmatists of the time drew upon developments in modern science to assist their thinking, particularly the theories of Darwin and Einstein. In following their lead and including our discussion of ethics previously, we too need to consider understandings that have accrued from quantum mechanics, microbiology and artificial intelligence involving outcomes such as intelligent robots, the genome project and more recently the 'Human Cell Atlas' (2017) project, designed to map every single cell of the body. Mead's writing about relativity, space-time and adaptation was the response of pragmatism to where human thought originates and how original, novel and ethical ideas are created by everyone; this is the question of 'emergence'. He wrote in a dialectical manner, when he described animals and humans existing in different systems (such as bees noted above) and how this needs to happen 'at once'. He discusses the

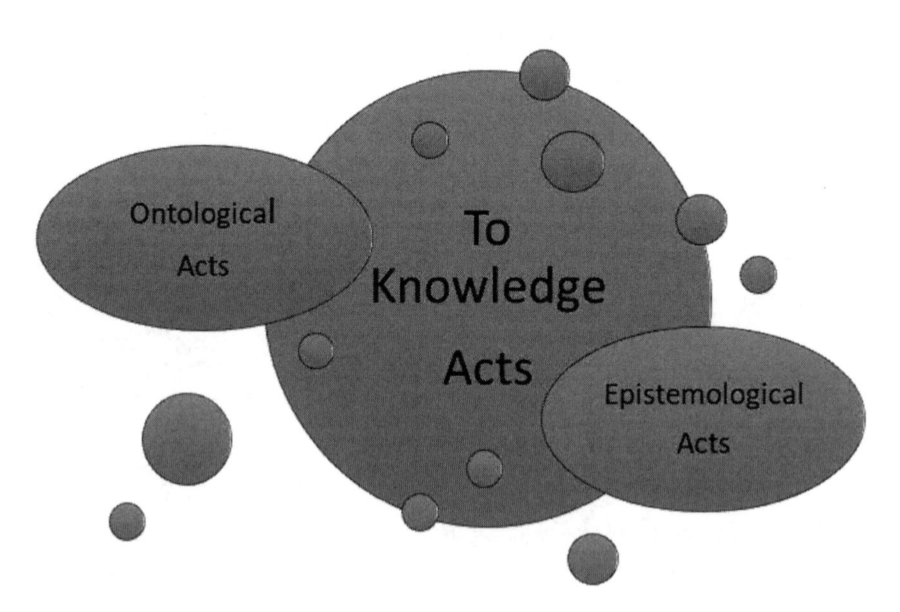

FIGURE 5.1 Representation of 'to knowledge' as a process of social action

emergence of consciousness and I propose subjectivity as feeling, in this evocative way (Mead, 1932/2002, p. 92):

> However I have defined emergence as the presence of things in two or more different systems, in such a fashion that its presence in a later system changes its character in the earlier system or systems to which it belongs. Hence when we say that the lowest form of consciousness is feeling, what is implied is that when living forms enter such a systematic process that they react purposively and as wholes to their own conditions, consciousness as feeling arises within life.

I have called what Mead is discussing here as the 'intersubjective moment', when a 'feeling' of change to the organism becomes apparent, when we become aware of our own existence and histories in the situations we confront. Mead says that this feeling occurs in the present, although we are always bringing together our experiences from the past and looking to future social acts that are in our interests. Time has been one of the great philosophical problems throughout the centuries and Mead's theories of simultaneous existence in different systems is designed to take this into account. I suggest that Mead has successfully integrated notions of adaptation from Darwin, relative positioning of ideas and experiences in different systems or conditions of living from Einstein and a dialectical biological process from Hegel and Marx to describe the very source of our humanness. Figure 5.1 is my attempt at depicting this source very generally. In my view, we have arrived at a point in our discussion where it is appropriate to accept that pragmatism can explain in descriptive terms, how thinking is initiated and proceeds with the evolution of new ideas that are novel, fresh and unique without need for other stimulus except matter and energy. It may be that we should add information and knowledge (Davies, 2008) and consciousness (Chalmers, 1996) so described as basic elements of the universe as well, properties of matter and energy. Why we should be able to create the novel and new is also answered in evolutionary terms, given that this capability enables humans to interact with all facets of their environments and systems for safety and progress. But we should also consider whether one of the great achievements of human kind, that of pure and applied mathematics, contributes in any way to our humanity, or not.

Mathematics and humanity

I have struggled with the philosophy of mathematics for many years, given my experience as a former teacher of mathematics and science at the secondary level and my quest to make the key concepts of mathematics doable and accessible for all children. In this regard, I have attempted to learn from other perspectives such as Ethnomathematics and the Indigenous (Hooley, 2010, pp. 215–232). Throughout this time, I have also been attracted to the idea of 'ambiguity' in mathematics, defined by Byers (2007, p. 28) as 'a single situation or idea that is perceived in two self-consistent but mutually incompatible frames of reference'. Whether or not Byers was aware of Mead's approach discussed above I don't know, but there is striking similarity. At a somewhat simplistic level, school mathematics can certainly be conceived as operating within different frames of references or systems according to students, teachers, parents, curriculum designers and economists. Rather than recognising ambiguity in how different groups understand the world, schooling attempts to enforce one view of how social practice should be conducted. To make this point, we can note Chesky (2014, p. 9) who cites Ernest (2004, p. 6) in his identification of five discreet aims of mathematics education (Table 5.2).

It can be envisaged that each approach to the aim of mathematics in schools will lead to different pedagogies being adopted. For this purpose, Chesky combines the first two aims into what he calls utilitarian, Old Humanist is relabelled cognitive and the final two aims he calls democratic. However Chesky makes a more fundamental distinction of themes within mathematical discourse for schooling when he identifies the following three philosophical groupings:

- Axiological: Utilitarian, Cognitive, Democratic (policy aims for mathematics education).
- Epistemological: Traditional, Constructivist, Transformative (pedagogies of mathematics education).
- Ontological: Absolute, Fallibilist, Aesthetic (conceptions of mathematics).

This formulation connects strongly with our discussion to date and the dialectic shown in Figure 5.1. It brings together ontological and epistemological acts of social practice and includes axiology, taken to mean philosophical studies and investigations of value, ethics and aesthetics. Introducing the concept of aesthetics at this stage not only demands that we give serious consideration to art, beauty and taste, but that we see mathematics in this light as well. That is, mathematics is concerned with the patterns, interconnections and relationships between objects, not so much the objects themselves, giving rise to attitudes towards such

TABLE 5.2 Aims of mathematics education

- Industrial Trainer aims – 'back to basics', numeracy and social training in obedience (authoritarian)
- Technological Pragmatists aims – useful mathematics to the appropriate level and knowledge and skill certification (industry-centred)
- Old Humanist aims – transmission of the body of mathematical knowledge (mathematics centred)
- Progressive Educators aims – creativity, self-realisation through mathematics (child-centred)
- Public Educator aims – critical awareness and democratic citizenship via mathematics (social justice centred)

objects. We appreciate the beauty of frost patterns on the windows because of the angles, lengths and curves involved taken as a whole not necessarily because we value snowflakes themselves. In other words, mathematics becomes another cohesive process for our human connections with the world and how we see ourselves. Aesthetics takes its place alongside an absolutist view of mathematics where mathematical truth has an objective existence outside of the human subject and a fallibilist view where mathematical understanding emerges and is evolving from human experience, culture and history. These are perspectives of mathematical knowledge, but can be views of knowledge generally as well, that is interpretation of the human organism.

Physicist and cosmologist Max Tegmark has theorised ideas similar to these in his consideration of the nature of human consciousness. He conceives of an external reality to ourselves that we know by mathematical description, an internal reality that each of us knows through subjective experience and a consensus reality that we agree and accept as a shared description subjectively and in terms of mathematics and physics. Tegmark puts forward what he calls the External Reality Hypothesis that proposes an external physical reality completely independent of human experience and the Mathematical Universe Hypothesis whereby our external physical reality is a mathematical structure (Tegmark, 2014, p. 254). In coming to this position, he notes that all theories we have about the universe are expressed in words and equations, symbols that we have invented. He calls these symbols 'baggage' and suggests that non-humans, aliens or a supercomputer could work on problems of the universe in different ways without recourse to idiosyncratic human 'baggage'. It does beg the question of whether these other beings or machines would come complete with their own incumbrances. In stating that an external reality independent of human experience means a physical reality that is a mathematical structure, he concludes (*ibid.*, p. 260):

> Nothing else has a baggage-free description. In other words, we all live in a gigantic mathematical object – one that's more elaborate than a dodecahedron and probably also more complex than objects with intimidating names such as Calabi-Yau manifolds, tensor bundles and Hilbert spaces, which appear in today's most advanced physics theories. Everything in our world is purely mathematical – including you.

These ideas are difficult to grasp and can be easily discounted. But I think that is exactly the point, to think about ourselves in the most thorough-going way and to then connect ourselves with daily practice as best we can. It is on this basis that our views of human purpose and direction are established and implemented in institutions like family, schooling, community and government. Tegmark's theorising of the mathematical universe and the mathematical you will confound and stimulate human meaning and how such issues can be pursued in democratic and formal education. What I now want to do is to describe my own mathematics classes that at least provide the opportunity of coming into contact with

SCHEMA 1: FACEBOOK POST, 18 AUGUST 2017

Having raised some worries about how school mathematics is generally conducted from a knowledge or epistemological perspective, what are some realistic changes that I would now recommend? My proposals that follow apply to learning generally in any school subject, but are especially designed to enhance language and mathematics for all participants regardless of background. I will draw primarily on the theorising of Freire and his notion of culture or learning circles and that of Papert in relation to his concept of mathetics (knowledge of learning, from Gk mathmatikos 'disposed to learn'), that links learning generally with mathematics specifically. As I have been arguing (for many years), school mathematics needs to be rethought as an open-ended, creative and cultural philosophical investigation of the social and physical worlds, not a mere application of known formulaic knowledge or procedure. It needs to emphasise non-imposed personally meaningful practices that establish new relationships and representations of knowledge for each person concerned. Next, groups of researchers are involved in experimenting with or building artefacts or objects of knowledge so that their different properties can be directly experienced. Third, learning circles immerse all participants in expansive language experience where ideas are explored, refined and communicated as widely as possible with regular public exhibition. Ideally this process should take as much time as necessary for ideas to be formulated and not hurried, but this is a major barrier to legitimate inquiry.

What I should do now is to provide a few hundred examples that can be taken up at various levels – but let's begin with one unfortunately brief broadly based on my own Year 7–10 teaching. My maths and science classes were usually arranged in small groups or learning circles of four to six members each. I attempted to have groups work outside whenever possible, to utilise a community general purpose workshop whenever that was available and, at a later stage of my teaching, to incorporate computer-based activity when I had programs that were appropriate. Colleagues were constantly compiling new units and projects that could be used under these arrangements supporting independence of thought and action. Investigating plants and animals that lived nearby in gardens, plants and creek provided obvious starting points for study and documentation. It was possible to collect specimens, compare and contrast structures, name features, think about families of items, check thoughts in the library or Internet or in the kitchen at home with experts such as brothers and sisters, raise questions about purpose and evolution. In the workshop, we were able to make mosaics and/or collages of specimens, photographs and graphics to convey knowledge and explanation of our specific study. Discussions and sketches could be made of shapes, relationships such as size and number of petals, construction of wings, body shape with brief informal notes. This type of semi-autonomous project work could occur over a period of weeks concluding with a small or more elaborate exhibition. Project materials become available as the starting point for other groups at another time. My question for you dear reader, is whether you see language and mathematics (mathetics, philosophy, philology) embedded in this process and whether it can form the basis of systematic and rigorous learning throughout schooling. We should consider the role of the teacher in this and the incorporation of reading, other materials and other experiences for instance excursions in addition to the sketch above. As far as I can see, language and mathematics are everywhere to be thought, experienced and lived. Neil

mathematics as a philosophy of practice and generate thinking in a dialectical transformational manner. I shall attempt this via a schema or informal social media post to colleagues as part of a communicative series on mathematics and the nature of school mathematics.

When I began this narrative I had not expected that the question of human subjectivity would take us to cosmology, theoretical physics and mathematics. What I have discovered is that 'everything is connected to everything else' and that as soon as my thoughts take me to one place, they quickly move on to many others as well. This seems to be a key principle of pragmatism. I realise that Figure 5.1 does not describe or explain the actual process of 'to knowledge' through 'dialectical emergence', but I think that it shows the relationship or alignment of ontological and epistemological acts that generate something else of the human condition that we can call subjectivity or consciousness. This occurs through a process of 'dialectical emergence'. Since previous writing (Hooley, 2015, pp. 180–181), I understand that major advances in the nature of human consciousness have not been made, but links with current major ideas in science such as quantum mechanics continue (Stapp, 2009). It is in this sense of our best understandings to date that we need to ask what is significant about our humanness that makes this journey imperative, why do we want and need to understand what lies at the centre of our humanity and do we need to incorporate all historical and cultural experience to advance this cause? In Figure 5.1, I have graphically shown a dialectical relationship between ontology and epistemology, but I have not attempted a description of possible changes to ontology and epistemology themselves as part of this process. This is required in line with dialectical materialism. First, is it possible that human ontology or human being has changed over the centuries in the way that human physical form has evolved? I suggest that the list of ten human characteristics developed in Chapter 2 have remained the same, but their application to problem solving will have been modified as conditions altered. In relation to this first question, the duality of human epistemology as content and as process will have also remained the same in structure, but will have altered considerably in application given the major economic, political and cultural changes that persist. There is no problem in envisaging each aspect of the dialectic undergoing a different change process over long periods of time while maintaining essentially the same relationship. What this means is that our view of humanness also remains the same, with all humans having a mutual interest in peace and harmony for evolution of the species. Human subjectivity and consciousness will in the end prove too strong for war and aggression, our place in the universe secure.

Dialectical emergence and knowledge duality

Following the educational notion of subjectification as outlined by Biesta, we have been investigating the origin of human subjectivity and intersubjectivity. These are difficult terms to define and involve the concept of human consciousness as well. However they provide an important basis for all educators to consider the characteristics of the classes they face, not in an essentialist sense, but as a framework for social practice. I have suggested that there is a constant interchange between ontological and epistemological processes to generate thought and have termed this 'dialectical emergence'. While it is not possible to describe the detailed nature of 'dialectical emergence' processes in the human brain, we can refer to the major ideas of science regarding quantum mechanics and microbiology for guidance. Until more scientific data becomes available, we continue to rely on broadly descriptive models with metaphors taken from current science such as ideation via 'crystallisation', 'isomerisation' and the like. In relation to knowledge itself, I have proposed that we view knowledge – or more correctly, 'to knowledge' – as a quantum duality, sometimes existing as 'structure' or content and sometimes as 'motion' or process. These considerations then take us back to Figure 5.1 and whether we can now refine the diagram to include these ideas.

As we saw, Mead proposed that humans can exist in different systems simultaneously which means that in the process of contradiction, that a human being is at each moment itself, but not itself, in transformation to another state. In this way, objective conditions are reflected in subjective thinking, or that external conditions give operation to internal conditions and the creation of new thinking. It is interesting in this regard that Mead refers to the Russian physicist and mathematician Hermann Minkowski (1864–1909), a former teacher of Einstein and who developed the notion of four-dimensional space-time. In discussing the nature of reality, Mead (1932/2002, p. 36) writes:

> The possibility remains of pushing the whole of real reality into a world of events in a Minkowski space-time that transcends our frames of reference and the character of events into a world of subsistent entities. How far such a conception of reality can be logically thought out I will not undertake to discuss. What seems to me of interest is the import which such a concept as that of irrevocability has in experience.

Here we grapple with the creation of the new or novel, again. Mead is attempting to show that events that occur and pass into our history do not remain unchanged or irrevocable, but can be reconstructed as we consider situations in the present. His concept of the past, present and future occurring at once. This requires different frames of reference where events continue to subsist, but provide the basis for reconstitutive practice. In pursuing this line of thinking in a quite remarkable piece of writing, Mead (*ibid.*, pp. 37–38) summarises:

> It is within this process that so-called conscious intelligence arises, for consciousness is both the difference which arises in the environment because of its relation to the organism in its organic process of adjustment and also the difference in the organism because of the change which has taken place in the environment. We refer to the first as meaning and to the second as ideation.

I take this passage to be based on Darwin's theory of adaptation, but extended past the physical to encompass the motion and transformation of human thought. The influence of dialectic and contradiction can be clearly seen. It is a purely descriptive explanation to locate meaning and ideas within this process, but I suggest it remains the best description we have. I now add human subjectivity to this process as well, depicted in Figure 5.2.

What I am suggesting with this diagram is a holistic and integrated 'being-in-the-world-ness' whereby humans have evolved to the extent that the constant dialectical interplay of social acts results in the constant emergence of distinctive human characteristics. It is probably the case that our language (another human characteristic to be discussed in Chapter 7) is currently inadequate to describe our own characteristics in the detail required and this may always be so, but we proceed within those limits. Heidegger (1889–1976) was a German philosopher who studied the question of what it means to be human and is best known for his book, *Being and Time* (Heidegger, 1962/2008). In this regard, he developed the concept of 'Dasein', meaning 'being there, presence, existence' that needs to be considered through progressive cycles of interpretation. I am not sure whether it is entirely necessary for all educators to delve into the deeper meanings of human

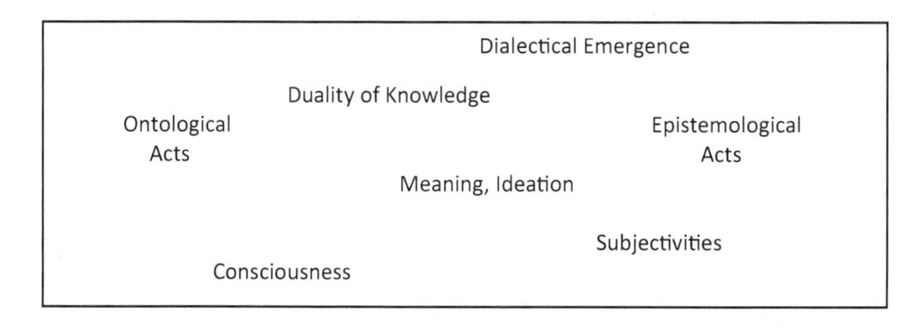

FIGURE 5.2 Representation of human subjectivity as process of social action

being that a reading of Heidegger would demand, but in reference to our guiding question of 'what does it mean to experience mind?' some attention to 'Being' is required. For our purposes and based on Figure 5.2, I make my own proposition that:

> Human being involves arrangements of matter and energy of the organism such that inner states and systems of awareness regarding past, present and future existence enable the sociality of dialectical materialist life.

We now come to the issue of 'how we are', of how we establish and nourish human being as just noted. Biesta (2013, pp. 17–19) discusses 'subjectification' as 'an interest in the subjectivity or subject-ness of those being educated – that is, in the assumption that those at whom our educational efforts are directed are not to be seen as objects but as subjects in their own right; subjects of action and responsibility'. As reported by Wyness and Lang (2016, p. 1053), Fielding (2006) makes an important distinction when he compares the conservative notion of school as a 'high-performance learning organisation', with school as a 'people-oriented learning community'. It is this latter approach that seems more likely to support 'the nurturing of children as good citizens and centred human beings'. From the discussion above, the process of subjectification can proceed as a result of dialectical emergence with knowledge existing as structure and motion. This can be considered as occurring in a constant, two-way process, from accumulation of knowledge of individual and particular events, to knowledge in general and, once a generalised view is achieved, to use this as a guide to encountering the individual and particular, including those that have not been encountered before. It is extremely difficult to account for the impact of certain experience on some people that can be quite different to the impact on others. But, if the interwoven nature of all experience could be mapped for each person, their viewpoints arising might be obvious. For example and perhaps one of the most difficult examples, those who have strongly racist views appear to be impervious to logical discussion and the presentation of evidence, indicating that their thinking is fixed and cannot be reconsidered, or 'dislodged'. Based on our argument of contradiction and its movement, this could mean that they have not had the opportunity of constructing their generalised idea from extensive experience of particular incidents, but have moved immediately to a simplified view that dominates. Human knowledge is not seen as a duality involving structure, motion and movement between the two. They do not appear to be interested in getting to the root-cause of issues through experience and then studying as many instances as possible to see that root-cause or essence in application. This is the basis of dogma in all fields, where viewpoints are accepted without question regardless of personal experience and evidence. A classroom example familiar to all mathematics teachers is that of fractions. This topic is usually taught in the same way around the world and is repeated in a number of year levels when test results indicate that understanding should be strengthened. Rather than repeating previous attempts, teachers need to provide

SCHEMA 2: FACEBOOK POST 14 OCTOBER 2017

Note. NAPLAN refers to a mass testing regime in Australia entitled 'National Assessment Program Literacy and Numeracy', for all students at Years 3, 5, 7 and 9.

Should we be worried about computer scoring of student writing, as will be trialled for NAPLAN in the near future? After all, if mass testing is unethical, inaccurate, misleading and doesn't tell us very much about student capability, why worry about any new proposal? How many teachers rely on NAPLAN to provide information about their students? Weighing the pig and chicken as many times as we like doesn't predict how we will enjoy the taste of our breakfast. That is, assessment and grading does not impact on the qualities of the artefact, the artefact stays the same no matter the assessment. Similarly, Mona Lisa stays the same regardless of the erudite critic. However let's consider the NAPLAN issue. It is quite possible of course to collect thousands of examples of say Year 9 writing and, on balance, summarise that there 10 adjectives per every 200 words of expressive writing. We can then collect those adjectives that are used the most and collect then in a computer data base. It would be very simple then to have a computer scan Year 9 writing and if 9 adjectives are found, allocate a grade of 9 out of 10 for every 200 words. A second data base could be compiled for words that are not used that often and another scan made and a score calculated. But what about creativity? No worries. Over time, the computer could identify words that do not fall into the first two categories that are then given to a group of expert teachers to classify. So now we have a third data base, those words that are judged by teachers to add expression: perhaps 'squishy' for watermelon rather than 'red', perhaps 'awesome' instead of 'good', perhaps 'bonzer' rather than 'ordinary'. We could generate a number of data bases like this, over time, for all parts of speech that it is considered generally characterise every 200 words of Year 9 expressive writing, or any type of writing at any year level for that matter. Scripts are then fed through a fast computer that checks for these arrangements of words, makes adjustments to account for irregular combinations (a common approach in statistics, adding constants, including factors) and calculates a score.

Now, I hear you say, what about human feeling, irony, pathos, humour and the like? Surely there is no trouble here either. Merely go through a similar process, have teachers assess as much writing as they can for these features, identify characteristics as best they can, develop the data bases accordingly. Gasp, is it possible that human feeling, irony, pathos are not that difficult to discern after all and a close reading can pick up on common patterns? Over time and constant checking and refinement by teachers, our

> data bases would become more sensitive to human expression and accordingly, more accurate in assessment (based on the assumptions made). We do have to ask difficult questions like how do teachers make their judgements in assessment and grading anyhow and how accurate are they? It may be that two or three different teachers or lecturers will assess and grade student writing in very similar ways (moderation), but they could all be inaccurate in relation to what the student is saying. This is the world of artificial intelligence that the military, business, medicine, law and many other areas have been incorporating for years to one extent or another. In a coming post, I might explain my view about NAPLAN and the possibilities of AI. I shall write to the US pentagon for advice. Neil

a range of different avenues into the concepts and practice of fractions, such that students can move from the particular to the general and then from the general to the particular in as many cases as possible. As Biesta suggests, students need to be seen as 'subjects of action and responsibility' in coming to understand aspects of the world and the impact that the world has on them. In the post below, I write to colleagues about another complex idea, that of artificial intelligence (to be discussed later) as applied to computer assessment of student writing, a development that is new and controversial in Australia. Evaluation and critique of the process will involve movement between the particular and general for a non-dogmatic appraisal. My suggestion at this stage and to be taken up in Chapter 6, is that the process of subjectification is central to our consideration of the evolution of human being and that the aspects of dialectical emergence and contradiction provide guidance for enabling the 'how' of human social awareness and sensitivity to be achieved.

There are difficult and new ideas and practices that we must confront every day, issues that are unfamiliar and which harbour potential danger of various types. Recognition that all humans must do this regardless of their socio-political background, culture, language and community experience is the basis of educational and social equity, cohesiveness rather than division. It also means that we exist within uncertain frameworks of subjectivities and subjectification as we relate to what is real and unreal. What we identify as human consciousness embraces this provisional totality.

Case 5. Learning from the land

'What are you doing for the break Michelle, off to the beach again?' Steve was his usual bundle of energy as he came into Michelle's office, always ready for a chat about what was going on in the faculty. 'Well probably,' replied Michelle, 'you know I always like to get out of the city'. They enjoyed many discussions about 'blowing away the cobwebs' and getting out into the fresh air on a regular basis. Michelle had grown up in a country town and was the first in her family to go to university. She decided to study biology and ecology and had worked for a while as a museum technician looking after the nature exhibits. When she had been asked to take up a part-time position in the science department of a regional university, she jumped at the chance. 'You know Steve, I was thinking, you know what you said the other day about having trees as your friend and I think I know what you mean.' They had been talking about where to take their science classes on their annual excursion and Steve had suggested a trip to his home country that was a volcanic area with craters, lakes, bushland and a variety of snakes. 'I can visit trees that have been there for as long as I can remember, I have photographs with my mother.' Michelle knew that Steve was a little sensitive to talking about the environment, especially as he had said a number of times that as an Aboriginal man, he had a special relationship with the land. He often used the expression, 'from the land, to the land', to describe the Indigenous connection and pointed out that if the land was taken away through colonisation or industrialisation, then death would follow. Michelle found this different or special connection with the land difficult to understand, because as a scientist and ecologist she thought that everyone had connections of some sort. Steve had a fine sense of humour, but he usually took on a serious look when this topic arose: 'Our community comes from between the rivers and stories are passed on from Elders year-by-year as the youngsters understand more. We know every tree, every rock, every animal, every billabong and it is our duty to protect them.' 'What does it feel like Steve to connect with a particular tree or rock,' Michelle said quietly, 'Is it like sitting in the shade of a big gum tree on a hot day, when the earth is trying to rest and cope with the heat?' Steve frowned and looked out the window as he gathered his thoughts: 'I don't think I can describe it to a non-Aboriginal person, someone who hasn't lived our culture and community, but it feels peaceful, like sitting with an old friend around the campfire.' He paused, then added, 'The land is not like family, the land is family and we feel close all the time, a part of us.' Michelle and Steve sat in silence, each cocooned with their own thoughts and experience. She respected Steve's views and the fact that he trusted her with sharing what he felt, deep down. It was doubtful if she would ever fully understand what he called 'Indigeneity', but she could appreciate that they saw things differently. 'Anyhow, this time next week, I'll be surfing, you should give it a go, you don't know what you are missing.'

6

THINKING ABOUT CONSCIOUSNESS

> But quantum mechanics raises other questions
> whereby all particles or potentialities are inseparable
> instantly correlated across universal vastness
> counterintuitive perhaps to the speed of light
> although when looking deep within another's eyes
> mysteries of truth and beauty reveal simultaneously.

Neoliberal accountability mechanisms such as national and international testing of school students have produced much debate around the world, particularly in regards the relationship between test results and socio-economic background. It is suggested for example that lower socio-economic background students have lower test results than higher socio-economic students. As a consequence, a 'deficit' theory of learning occurs related in particular to the level of family income. Placing aside for the moment the purpose and nature of neoliberal testing and connections it may have or not have with the culture of specific groupings of citizens, there is an epistemological argument to be made regarding the nature of knowledge and whether capitalist or neoliberal schooling takes account of how the vast proportion of the human population goes about learning. From the perspective of pragmatic philosophy, human learning takes place when unexamined and unreflective assumptions, ideas, predispositions and the like come into contact with new situations. As the human proceeds to grapple with the relationship between current understandings and possible developing thoughts, concepts, ideas and practices to resolve resulting tensions, new practices are created that enable progress and change.

In general terms, modern schooling does not attempt to monitor or track emerging practices, but instead, imposes often disconnected and superficial tasks that are supposed to illustrate the outcomes of learning, rather than learning itself. This contradiction raises the issue of human consciousness, the capability of being aware and reflective of previous personal and community history and experience. Sometimes called the 'hard issue' of philosophy, the problem of human consciousness pervades education, teaching, learning and assessment and, while denied, distorts terribly our understanding of learning outcomes for all children. If it can be argued that human consciousness is produced pragmatically as other thoughts and ideas, then how students reflect on their experience from the perspective of their previous experience and culture – broadly, their personal consciousness – will direct their own thinking and new concepts of the realities they face. A practice approach to understanding human experience and learning therefore does not take us down the path of 'deficit' theory, but instead opens up a view of humanity involved in a complex array of social practices that we engage together, as communities and through which we reach broad social consensus on our understanding of what is important, how communities develop and how the social and physical worlds interconnect.

Marvin Minsky (1927–2016) was an esteemed American cognitive scientist who worked in artificial intelligence and many related computer and philosophical fields. He began his book, Society of Mind (Minsky, 1986, p. 17), in the following manner:

> This book tries to explain how minds work. How can intelligence emerge from nonintelligence? To answer that, we'll show that you can build a mind from many little parts each mindless by itself. I'll call 'Society of Mind' this scheme in which each mind is made of many smaller processes. These we'll call *agents*. Each mental agent by itself can only do some simple thing that needs no mind or thought at all. Yet when we join these agents in societies – in certain very special ways – this leads to true intelligence.

Minsky has introduced the term 'intelligence' for the first time in these chapters, reflecting of course his research into artificial intelligence, generally understood as computer outputs similar to human outputs. In this sense, it is relatively straightforward to see human behaviour being mimicked by the combination of numerous computer systems (see Schema 2 above), programs or *agents*. Our emphasis so far has been on epistemology and the nature of knowledge, as distinct from intelligence, but we shall link the two by considering intelligence as the capability to accumulate information and knowledge for application and problem solving within an environment. This working definition makes clear the case for both human and machine intelligence or outputs such as poetry writing, correcting essays, cleaning the floor, operating equipment, playing games and the like. Further comment on the nature of intelligence will occur later, as it is important for our purposes of education to have a view on how problem solving actually works: how to I decide to pick up that cup of coffee, to put one foot in a certain position to avoid a puddle of water, let alone to choose my words carefully when discussing a difficult issue? These are certainly difficult questions for how we understand human being and the role of knowledge. However, at this stage, I am not sure whether it is entirely necessary for all educators to delve into the deeper meanings of human being that for example a reading of Heidegger would demand, but in reference to our guiding question of 'what does it mean to experience mind?' some attention to 'being' or 'to be, to exist' is required. For our purposes, I repeat my definition of human being and then propose a more detailed note about knowledge:

> Human being involves arrangements of matter and energy of the organism such that inner states and systems of awareness regarding past, present and future existence enable the sociality of dialectical materialist life.

A central aspect of human being arises from the infinite and continuing process of 'knowledge knowing itself' whereby the dialectical duality of knowledge as structure and motion creates an awareness within the organism of what it is like to be like something else.

I have taken the key concept of 'knowledge knowing itself' from Hegel (1977, p. 400) as a means of thinking about the production of human feeling from knowledge. When I eat an apple, I have content knowledge of that object as a fruit grown on a tree and subjective knowledge of what that object is like from my sensory engagement and responses. The experience of this dialectic duality creates in the human organism a sense, or feeling, or awareness of what an apple is like. Human feeling or awareness then can be considered as an outcome of the knowledge qua knowledge process, or as a materialisation of a bodily entity, of another form of knowledge. In this way, we are coming closer to the idea again that human subjectivity, consciousness or feeling arises from a materialist process in the body that is taking place constantly from birth to death, during every moment of life, without the need to propose a metaphysical dimension of some type. As I understand it however Hegel was concerned not so much with complete insight and understanding of the universe and ourselves, but with human appreciation of the awareness of being, of what it feels to be human, as part of the absolute or wholeness. From my point of view, I think the two go together as we act and live with and in the world and through that action are taken to broader perspectives beyond the horizon. This can certainly be the imperative of formal education in all economies and cultures informed by a reconceptualised understanding of knowledge (Figure 6.1).

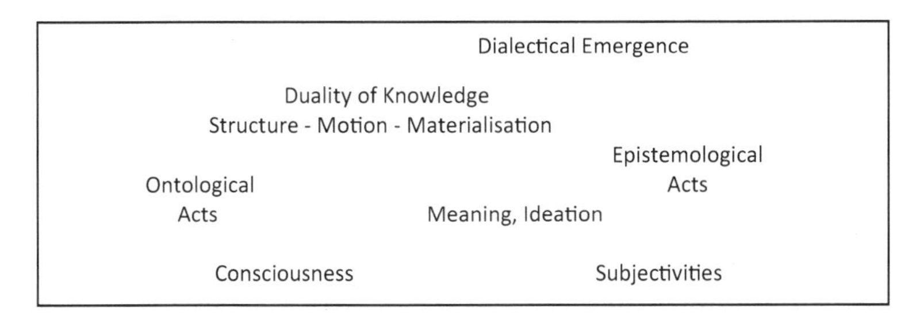

FIGURE 6.1 Representation of 'knowledge qua knowledge' within process of social action

SCHEMA 3: FACEBOOK POST 26 NOVEMBER 2017

Education is one of the great delights and virtues of life, perhaps the greatest. It constantly fills our hearts with anticipation and our minds with wonder. From the moment we discern the natter of birds in the morning, to the gathering of vast expanses of cloud in the afternoon, to the faint hues of light gradually fading at dusk, we chart the passage of personal growth and feeling. Throughout the day our relationship with the world has altered as we grasp sound and sight such that they themselves will never be the same again. Who knows how it works, a touch, a glance, a kiss that connects with others, a mother's soothing word, a friend's tone of doubt but expression of support. Education is how we understand the interrelationships of all entities and how we bring past, present and future together in judging how to proceed in the present. Education is recognition of the great beyond, based on our experience that imagines new ideas of what might be, a better existence for all. Education enables the organism to feel what it is like to be ourselves, but also something else and to exhibit indignation, sadness and compassion. A child is perpetual motion transforming matter and energy with imagination and with each new encounter. An older person observes with patience as the coming generation must explore and understand for itself, unexpected outcomes and puzzlements the stuff of curiosity. We have come to this position, of 'how we are', after many centuries of meeting the unknown, noting the effects of what happens and making the best estimation of next steps. This seems reasonable, even to where what we value and how we should act properly originates, but what we 'feel' remains another matter. Or at least, the same matter, but generating something different, our subjectivity. There is no need to imagine human subjectivity as something other than the sensitive workings of matter and energy, although it will be some time before those actual workings are identified. What we can say is that our subjectivity, our feeling of humanness when actively connecting with all the entities, cultures, histories, languages and uncertainties of life, we each know and find precious and honourable in those close in heart and mind. Education is characteristic of human being and common to all. Neil

Intimate feelings of being

John Searle (1932–) the American philosopher does not have too many problems with the nature of consciousness. As I recall, he does not agree with Chalmers that consciousness is the 'hard problem' of philosophy and instead writes confidently (Searle, 1998, pp. 40–41):

> The primary and most essential features of minds is consciousness. I mean those states of sentience or awareness that typically begin when we wake up in the morning from a dreamless sleep and continue throughout the day until we fall asleep again. Other ways in which consciousness can cease is if we die, go into a coma, or otherwise become 'unconscious.' Consciousness comes in a very large number of forms and varieties. The essential features of consciousness, in all its forms are its inner, qualitative and subjective nature in the special sense of these words.

This statement does not exactly explain what consciousness is, but puts forward three features of its presence. Searle suggests consciousness is inner, because it occurs within the body and brain of each person, has a qualitative awareness of what it feels like to be something else and is subjective in that feelings are experienced by a human subject. These are important ideas, but a description of an apple as having a red or green outer skin, an inner flesh with juice and a core with seeds, does not really enable us to 'feel' what it means to be apple. Nor does it explain the dialectical relationship between human and apple and how each are changed in motion. At this point then, we can reconsider our question of 'what does it mean to experience mind, to act, think, know and create ethically?' in relation to being, subjectivity and consciousness, in a little more detail:

- 'Mean' can be a relative term regarding the purpose of human being on planet Earth, or an absolute concept regarding the significance of human being in the universe.
- 'Experience' refers to knowledge gained through the social action of living in and with the world.
- 'Mind' is that function of the brain that enables feelings of awareness of entities of the world and of its intellectual capabilities to occur.
- 'Act, think, know and create' are all actions that arise from social engagement with the world and the process of 'knowledge knowing itself'.
- 'Ethically' concerns acting in the best interests of humans and humanity through active social engagement with other humans.

From this, we can reconstruct the question as:

- What is the significance of human being on Earth such that experience of living in and with the world enables practice and consciousness of worldly and social entities and intellectual capabilities?

What is the importance of a statement like this for educators? As a starting point into this question, let me describe how I see its importance for me. What do I think of myself, about how I have conducted my life and my connection with, or evaluation, of my social being at present? I really only see myself in two ways, that is as an Australian citizen and as a member of the international working class and all the connotations that both entail. I do not have a feeling of being masculine, of belonging to a particular race of people except the human race generally, of being Indigenous or ethnic. I imagine I feel Australian because of being born in a country town of Australia close to a distinctive bush and beach environment. I feel a member of the international working class I imagine because of the economic background of my family who always worked for a living rather than had any ownership of how goods and services were produced. I have no concept of being a male and I hope that men are kind and generous, love and respect their families and communities and do not wish to invade other countries for profit and annexation. As mentioned previously, I do not conceive of a personal 'self' where these qualities or properties are supposedly located, but see them as regular outcomes of the definitions above, through social experience and knowledge formation. Then again, do I feel human? This is an impossible question as, in accordance with Searle, there are no agreed comparisons to be made from inner, qualitative and subjective thoughts. Like happiness and virtue, I may believe that I feel human on fleeting occasions when at peace with the natural landscape after an angry storm, when spending hours with someone close talking about something or nothing, when watching a baby laugh and grow. Having some broad framework as we have sketched about 'being in the world', whether definitive, or somewhat vague, provides the educator with a point of reference for relating to those in each class and a philosophical direction for considering dialectical emergence.

I have used the word 'entity' above in the sense of an item that exists as itself, an object in thought or actuality. As such, this is an important concept in philosophy and in our discussion of consciousness. Dewey (1916, pp. 103–104) for example noted that:

> We are only too given to making an entity out of the abstract noun 'consciousness.' To be conscious is to be aware of what we are about; conscious signifies the deliberate, observant, planning traits of activity. Consciousness is nothing which we have which gazes idly on the scene around one or which has impressions made upon it by physical things; it is a name for the purposeful quality of an activity, for the fact that it is directed by an aim.

There is considerable similarity between Dewey's statement and what we have considered above. He does however make a clear distinction between our understanding of what we are as a noun and instead, defines consciousness as a 'purposeful quality' of an action. His comment about being 'directed by an aim' is not an after-thought but indicates the overall rationale of consciousness to resolve situations so that the human organism can grow and adapt in relation to the environment. Dewey extends his discussion to include a previous trend in epistemology that identified mind with 'self' and the use of the term consciousness as being equivalent to mind, 'an inner world of conscious states and processes,

SCHEMA 4: FACEBOOK POST 22 JULY 2017

How do we know what is real? Am I really looking at a koala, or is it a kangaroo? More to the point, do you agree with me that it is a koala, or do you see a possum? This is the question that has puzzled philosophers for centuries, the question of how the human brain (I am assuming brain for these notes.) 'represents' reality, whether reality exists in brains only (reality constructed by brains) and do different brains have different realities? However the brain needs to be able to distinguish between koala, kangaroo and possum in the same way that it distinguishes between Auntie Joan and Uncle Jack when they visit. It is tempting to compare this process these days to data processing and how a computer digitises objects (text, drawings, photographs, music, maps) and stores them somewhere ready for retrieval in their original form. We don't know the answer to these questions, but it seems we do need to be able to experience, transfer and transform that experience in the brain and then be able to draw upon that experience when we want to compare, contrast, relate, remember, evaluate, modify etc. To use the concept of Mead, the human organism must be very 'sensitive' to enable this to happen, a process honed by evolution. In terms of formal education, I think a number of fascinating issues ascend that are of interest to the educator, parent, family and friend: why does this child have a sense of humour and this one not, how does this person 'read between the lines' and this one not, why is one twin compassionate and the other not, where do thoughts of fascism come from? Whether we want to speculate on reasons here or leave them alone bears upon whether we want a defensible frame of epistemology as we approach each and every discussion with people who matter. It seems to me that when I walk into my Year 9 mathematics class I should have a definite way of thinking about the task at hand and how I can encourage every student to consider a range of difficult ideas and practices, unknown to them when they entered. I should be able to accept that the sensitive evolutionary organisms that await me are different one from the other and will act and respond accordingly. Those classic diversionary mechanisms of humour, feigned ignorance and confusion, interfering with others, claiming attention and the like may in fact be serious clues of creative thinking, of attempting to tackle the idea from a different perspective. If we do not understand how each of us engages what is real, then these trial and error excursions are exceedingly important. Viewing epistemology creatively, as involving avenues for original personal thought, means a different type of schooling. As a specific case, it brings us much closer to mathematical thinking than what is usually encountered in most schools around the world. Neil independent of any relationship to nature and society, an inner world more truly and immediately known than anything else' (*ibid.*, p. 293). This position would suggest that conscious states comprised of entities are preordained and internal for each individual independent of their personal interests, meaning that thought is constrained and not liberated. I am not sure if Dewey had a view on Aristotle's notion of entity that is different from his, but it is an important concept in philosophy. An entity is taken to mean an object that has a distinct and independent existence and enables us to think about the relationship between acts, thoughts, outcomes, possibilities and the like. Aristotle spoke of what he called 'first philosophy' as consisting of entities that have no physical existence, including the 'prime mover' of the universe and 'second philosophy' that involved physical entities. Metaphysics is inquiry into 'first philosophy', while physics is inquiry into 'second philosophy'. Separation between the two is often difficult to maintain. These are important philosophical considerations when we attempt to theorise how human understanding occurs and how connections might be made across the various although continuous stages of thought. Pragmatism had to deal with these problems if it was to provide a different explanation of human being to *a priori* and respect the capability of every single human to live knowledgeable, productive and dignified lives.

Problem solving process as human action

In quite a remarkable book, Mead (1938/1972) goes into fine detail regarding the character of the human social act and the implications that arise such as connections with the major trends of modern science, formation of the self, the concept of mind and many more. For example, he writes (p. 377):

> The very nature of the idea indicates that in a very real sense, it is not in the world. One's understanding is that of fashioning the world so that it shall conform to an idea that is not there realised and that reconstructed world is an environment that is confined, at least for the time being, to the individual whose idea it is.

Thus, ideas involve an interchange of entities. Through acting in the world, the human organism draws upon previous and current entities of practice and thought, is able to relate these in some way to what is presented and can judge and decide the most advantageous action in response. This process is the basis of consciousness and self. Mead notes 'Consciousness becomes our experience of things not as they are but as they impress us from a distance which we can never overcome except in imagination' (p. 74). He suggests that this approach to consciousness leads us to 'appearance of' (p. 75) or of awareness. I am more than comfortable with the notion of subjective awareness being a way of thinking about the problem of consciousness and that this awareness is a function of physical processes in the brain, the interconnections of matter and energy. This is a logical outcome of Mead's socio-biological view of pragmatism that, as commented on previously, was strongly influenced by Darwin's account of animal and human evolutionary development and the empirical basis of science. He locates emergence of the self in the same way, stating that when a person can take on the attitude of another and thereby forming a relationship between one who indicates and one who is indicated towards, then the person who is indicated towards becomes a social object along with the other social objects present: 'When the memory of the indication associates itself with this object, the self has appeared' (p. 75). A strict distinction between these concepts and their constituent entities is difficult, that is social acts, thoughts, ideas, self and consciousness meld together in the continuous integrative functioning of the human organism.

Within this paradigm of consciousness and awareness, how then do we consider the human characteristic of problem posing and problem solving, a question of supreme interest to educators at all levels? Let me first refer to a recent article (*New Scientist*, 2017, p. 9) that reported an artificial intelligence programme called AlphaGo Zero was beating all its human opponents when playing the Chinese game, Go. The programme did not begin with an extensive data base of human-derived rules about playing the game, but starts playing at random, against itself and storing moves in memory. Apparently, after 3 days of such activity and nearly 5 million games, it became the best Go player ever. According to the report and main researcher for the project: 'It's more powerful than previous approaches because we've removed the constraints of human knowledge' and that 'AlphaGo Zero was able to rediscover much of this Go knowledge, as well as novel strategies that provide new insights into the oldest of games'. To have a detailed analysis here, we need to understand how the programme works, but in broad terms, it does remind us of Minsky above, the notion of creating human intelligence from an integrated society of very small non-intelligent agents. The question of 'randomness' in human thought and meaning is also interesting, as it has been possible for many years for computer-based applications to respond randomly from available data bases in a 'sensible and understood' manner. It is indeed a significant comment above that sensible outcomes can be achieved by computer technology processes when the 'constraints of human knowledge' have been removed. This fits nicely with our discussion to date. I interpret this statement as saying that human knowledge defined as content or packages does not have to be present in abundance before humans can think and think creatively with new ideas. Rather, we are able to construct new thought and ideas through the interaction of the various entities noted above, that result in awareness of the world and of others. This is the meaning of Mead when he describes 'indication of and towards' that produces feeling of self. It is difficult to grasp these concepts arising from matter, energy and associated entities, but because there is nothing else, they act and we observe and feel their existence in every moment of life.

In discussing the question of consciousness, Tegmark (2017, pp. 284–286) provides a hierarchy of concepts that need study. He calls these euphemistically and based on the Chalmers categorisation, the easy, pretty hard, even harder and

SCHEMA 5: FACEBOOK POST 11 AUGUST 2017

Epistemology is a vexed question. Generally thought of as the theory of knowledge and learning, the nature of knowledge and its origin, where ideas and understanding come from, it has been debated by philosophers for centuries. There is the issue of how humans consider their sense data and whether we exercise reason on sense data for knowledge to occur. Some argue that 'justified true belief' is the basis of knowledge, but there are intriguing examples of how this can be contravened. What I want to raise here for educators is the problem of the (possible) existence of multiple epistemologies particularly those ascribed to different groups in society including Indigenous, feminist, ethnic, working class and marginalised. I can well remember for instance the notion of 'working class curriculum' usually thought of as being more practical than abstract and connected at the secondary level with employment. Can we however go a little more deeply than what is called 'ways of knowing' and consider where our thought actually 'comes from' or is produced? This question is also one that has puzzled us without a definite answer yet available. For me, we need some concept of 'consciousness' (including language) involving internal modes or structures or states that enable humans to interact with the external social and physical worlds through our sensual mechanisms, forms including touch, sound, taste, text, image, numbers, story, music, drama, poetry, ceremony, meditation.

One of the original American Pragmatists G. H. Mead used the concept of 'intersubjectivity' to help explain this process whereby current arrangement of the brain connect with new experiential structures to create original thinking or thought (called mentalise – there are many issues here that cannot be pinned down fully as yet). Indigenous knowing can help focus thinking on these matters and whether Indigenous thought occurs through a similar intersubjective process as mentioned above with emphasis on 'ways of knowing' involving contact with the land, storytelling, the wisdom of Elders and families, cultural practices and ceremonies and centrality of language. These issues can be found in the activities of most social groupings, but perhaps are more emphatic for Indigenous communities. There is often a distinction made between 'cultural' (seen as organic) approaches to learning and 'scientific' (seen as mechanistic) approaches, but I think this is a false distinction. Whatever community, personal or cultural activities are accentuated, the issue here is how brains work and as noted above, make sense of the inputs that we have. I cannot accept that the brains of the groups mentioned 'work' differently even though different activities are involved. Again, I am not suggesting that all children from all social groups must be forced into the same way of learning at school – a common criticism when epistemology is raised – and I strongly argue that epistemological frameworks in all classrooms need to be flexible and diverse. I am very pleased for example to work within a generalised inquiry framework of Indigenous 'ways of knowing' across the curriculum as I see many interconnections across these approaches. I suppose the key point I am trying to make is to propose that all educators need to have an epistemological frame of practice within which they and students live, act and learn, to recognise different 'ways of knowing' and to value the commonalities that exist at the structural level. Neil

really hard problems of consciousness. These involve how the brain processes information, why some arrangements of matter are conscious and others not, what gives specific matter its qualia and finally, why is anything conscious? He suggests that science is making progress on testing the initial questions and therefore in constructing a scientific theory of consciousness, but the 'really hard' problem remains. In similar vein to 'knowledge knowing knowledge', Tegmark comments that consciousness may depend not so much on the matter doing the information processing, but on the structure of the information processing itself (p. 315). This takes us back to motion and transformation being at the heart of the dynamic universe and at the heart of dynamic ourselves, described above as 'dialectical emergence'.

I can well-accept that the physics of the brain may be different to the physics of the universe, in the same way that we believe our current understanding of the laws of physics did not hold immediately after the Big Bang. However in a similar manner to Aristotle, Newton and Hawking, our present understanding and imagination including quantum mechanics and microbiology is what we have and our starting point. Theoretical physics and the weird world of quantum mechanics makes amazing theorising possible, but we need to combine the physicist and the philosopher to engage reality and to be able to think about phenomena and epiphenomea. It seems to me then that we have interactions occurring in the matter and energy of brain substance that result in organic feelings of what we call self, subjectivity and consciousness. Drawing on classical mechanics and quantum mechanics, I can envisage the function of brain substance being the dynamic, continuous formation of electromagnetic waves that collapse to singular points of entity awareness, the source of our humanity. It should be noted that neuroscientist Susan Greenfield discusses a similar notion of 'neuronal assemblies' acting in this way, populations of neurons connected to and informing consciousness (Greenfield, 2016, pp. 40–51). Feeling occurs because of the particular arrangement of matter and energy that has evolved over the centuries to constitute the human organism and its life force. Accordingly, everything else has a viable and reasonable explanation, or at least description of some type. Infinite acts of dialectical relationship between ontological and epistemological transformation produces 'knowledge knowing knowledge' and the creative dialectical emergence of various novel entities of thought and further action. Considering 'duality of knowledge' proceeding through contradiction of structure, motion and materialisation, that is recursive entity formation, attenuated or intensified depending on previous acts of experience, is a progressive and optimistic philosophical view of humanness rather than mere bourgeois psychological or sociological doctrine. Human values and ethical conduct are social systems and routines that are constructed and consensual over long periods of time and practice. This is a radical indeed liberatory view I think that has significance for all educators, as it accepts consciousness, self, subjectivity and autonomy (Wilson and Ryg, 2015) as human feelings and understandings of humanity that apply to everyone, every member of society regardless of background and every student in every classroom. It is a view that sees the human species as comprised of democratic, inquiry and integrated organisms. It is a view that focuses on the conditions of experience that supports all children to be aware of their own capabilities and how they can interact with the physical and social environments to change what they experience into something more meaningful and beneficial for all. It is not a question of individual, mysterious consciousness being bestowed from without, private and exclusive over which we have little control, but our existence as humans being constructed and collaborative, in societies of mind and action, for the historical and public good.

Case 6. My Lai Massacre

It was sometimes difficult to talk with Uncle Norm. Mum and Dad would say to Ken that he shouldn't worry if Uncle Norm seemed not to be in the mood to talk, or sat quietly by himself when visiting, as 'that's the way he is, he values his solitude'. In spite of this, Ken liked his uncle and sometimes when they went fishing off the bridge, it was just nice to sit without saying much, waiting for the fish to bite. Norm might ask occasionally how things were going at school and what books he was reading. Mum had said that reading was a passion on her side of the family and that her brother Norm had always read widely, something that he could do by himself without the interruptions of others. It was strange then that on a Saturday afternoon, Ken heard some raised voices coming from the kitchen. It was upsetting for him as his mum and dad never argued and Uncle Norm was known for his unflappable nature. Ken watched as his uncle slammed the back door and headed down to the end of their large back yard and to a wooden bench under a tree where he often sat by himself when visiting, 'to gather his thoughts', Mum said. Ken walked up to the bench, sat down and waited. After a few minutes, Norm said quietly almost to himself, 'It's hard to explain what it was like you know.' 'What do you mean Uncle Norm?' Gazing at some point in the distance, Norm continued, 'It's a long time ago now son, but war is a terrible thing. You know that I was conscripted into the army and sent off to fight in Vietnam, forced into the army and sent to a country most of us had never heard of, to fight a people we had no argument with.' Uncle Norm paused and Ken could see that there were tears in his eyes. 'It was some time after we got back home that we heard about a massacre of civilians and children at a village called My Lai, there was this American Lieutenant, William Calley who was in charge, over 400 were killed. He was put on trial, but only got a light sentence.' Ken had heard about the Vietnam War where Australian troops had been involved and he knew that the Americans had many troops, bombers and helicopters there. 'I just hated being a part of it. Your father missed out on being conscripted, so he doesn't know what it was really like,' Norm went on, 'and he has no idea what it's like to go on patrol with your mates, to have them shot, to not return.' Ken didn't know what to say as he could not imagine what war was like despite the films and documentaries he had watched and he could see that his uncle was very sad and distressed as he remembered what had happened, even many years ago. 'Thinking about My Lai brings it all back son, just like yesterday, awful memories. I can still feel the heat of the jungle, the thumping noise of helicopters, the first aid station in tents where they would bring the wounded, it was a different terrible world that I can't get out of my mind. I understand your father, but he doesn't understand me' Uncle Norm and Ken sat on the wooden bench at the bottom of the garden under the big old tree, keeping each other company as they did when fishing. It was enough they were together, related, connected by their thoughts and hopes.

7

LANGUAGE, THOUGHT AND CREATIVITY

> At dawn, the light of love emerges to behold
> in conflict with all that is divergent and obscene
> rebellious and obstinate in structuring the intuitive
> understandings of experience permeating the real
> as language and histories lay the basis of causation
> spreading across what we think is known and valued.

Strongly connected to the notion of human consciousness is that of language. Linguists and philosophers continue to debate the nature of consciousness and language given the role that each must play if humans are to inquire and create. Chomsky is of the view that language is a biological function of human growth in the same way that arms are produced rather than wings. This notion of language should not be confused with the rules and procedures that have been developed to govern reading and writing. Additionally, in one of his most famous statements, Wittgenstein (1922/2016, p. 108) proclaimed enigmatically, 'Whereof one cannot speak, thereof one must be silent.' This declaration has a number of interpretations regarding Wittgenstein's attempts at relating reality, language and meaning, all of which can be contested. For the purposes of this book, the statement will be considered as an attempt to show the difference between science and metaphysics, the former being concerned with facts and explanations and the latter with understandings abstracted from experience regarding the nature of being and of humanity. Wittgenstein's views of religion are a little obscure, but falling into the category of metaphysics, indicate an interest in contrasting viewpoints that are essentially scientific or metaphysical. Publication of the *Tractatus* in 1922 following the carnage of World War 1 and the continuing development of modern science, would necessarily raise questions regarding human purpose and being and whether answers to these problems and anxieties would more likely be found from direct experience and analysis, or from internal belief and thought. Wittgenstein may be saying that human perceptions and creations of beauty, poetry, compassion, sadness and of truth itself, do not come from strict analysis of experience of the physical world, but occur through human dispositions to embrace the world and each other. For example, the beauty of a rainbow is not taught through reading, instruction by experts or analysis of the colour spectrum, but occurs to us almost instantaneously as we observe a rainbow and its features. Walking into the ocean tells the young child about the properties of water even before there are words to connect. Feeling close to another person is often difficult to explain and words may seem inadequate, yet the feeling and emotion are known. Wittgenstein may be saying therefore that science seeks to describe, denote and explain in some areas of human life, but in others that go to the very depths of being human, we need to leave well alone. Could Wittgenstein be saying that science provides one path to meaning, religion another and that they exist in parallel? A neoliberal approach to education may not consider these considerations necessary, but for a pragmatic and intersubjective praxis, they are essential.

Two comments can be made regarding this interpretation of Wittgenstein. First, it is difficult to argue that instantaneous thoughts regarding the beauty of a flower arise without reference to previous cultural experience, indeed human experience that extends over thousands of years. A feeling of sadness when reading a story, or looking at a painting, does not occur in a vacuum, but within a context of human experience and activity involving nations, communities, families and individuals. It is possible to intellectualise where these feelings and emotions (see below) come from, but it is not possible to separate the various and interwoven factors of causation. It is also not possible to prevent humans in thinking about, discussing and investigating these outcomes either in private, quiet moments, or together in groups. In fact, when serious events happen that may cause collective delight or distress, people seek celebration or solace in each other as they express both satisfaction and support. That is, while feelings and emotions are difficult to speak, they do not remain silent. Second, in suggesting that science and metaphysics are fundamentally different, it is possible to argue that one should not dominate the other, a supposed problem that has been identified since the Enlightenment period. On the other hand, in posing this dichotomy, it separates two types of person, with two types of thinking. This stands in contrast to the concept of an integrated consciousness, capable of engaging a wide range of experience and problems, as required. Scientists may draw upon the extent of cultural and scientific understanding when considering their projects, especially in the planning stages of clarifying their intent, approaches to knowledge and research methodologies. The everyday work of scientists, technologists and engineers for example, may have close overlap with that of artists, politicians and priests as they seek to understand the essential nature of the tasks they face and in whose interests they work.

Language as act

In coming to the notion of language as tool, Wittgenstein focused on language use or act, rather than initial meaning. If language can be thought of as an inner state of being for humans that connects with the external world then, similar to consciousness, language is a sensuous human disposition that generates perception and conception of various types depending on social and physical conditions. As mentioned above, when snow falls, the young child does not need an extensive vocabulary to feel cold or to know the meaning of cold, or the wonder of softness and texture. Words will be added later to express these experiences and to communicate with others. Seen as a tool in this way, the inner state or mode of being will not separate causation, explanation, interpretation, feeling and the like, but will relate to the experience holistically. For the very young, it is difficult to understand exactly how perceptions produce specific thought when experience has not been had before, such as the first falling of snow. But the human organism will react based on the sum total of experience that has been accumulated to that time. In a similar way, humans will come to their own understanding of colour, sound and taste without words and without being taught by others. In conceptualising consciousness and language as inner states of being, the notion of 'language act' integrates other acts such as speech act and social act as humans pursue their quests for knowledge and understanding. We therefore conceive of humans as active, thoughtful beings, engaging with their inner and external worlds at all times from birth to death, coming to know and construct meaning both privately and publicly. Following Wittgenstein, we investigate the practice of language act and its place in what can be called 'intersubjective praxis' existence.

It seems to me that if our best scientific thinking (defined broadly as the search for causation and explanation of physical worlds) at present indicates that there are four fundamental interactions in the universe, then these will impact on human thinking and understanding. This is because the human brain/organism is constituted by the same materials and processes that make up the universe and is not separated from them. In other words, a materialist view of history and of humanity suggests that the characteristics of being human have evolved from the interactions of matter, energy, society and culture (all human concepts and constructs) and have not been preordained. What occurs within the human brain/organism involves gravitational, electromagnetic and strong and weak nuclear forces (all human concepts and constructs), but over the centuries there has developed amazingly sensitive and complex interactions and correlations between them. For these reasons, I see strong connections between what we call physics and metaphysics, rather than these being considered as two entirely discrete realms. One does not seek to dominate the other.

In suggesting that there is an area around a magnetic substance within which magnetic force is experienced, modern science has proposed the concept of field. A magnetic field for example is where each point of the field is influenced by a magnetic force. Similarly an electric field surrounds an electric charge that will act upon another particle that enters the field. Concepts like field and force help explain what we observe and consider real, although may still be difficult to describe and interpret in their own right. Given that the human brain/organism involves the four fundamental interactions mentioned above, it is possible to envisage infinite, interrelated and correlated epistemological fields being produced each nanosecond at the speed of light as we experience our social and physical worlds. From quantum mechanics, the notion of collapse of the wave function (a way of describing a quantum state in terms of position, momentum, spin) might occur, as many fields or states collapse to a single eigenstate of understanding. Drawing on some of the key concepts of physics and quantum mechanics then provides a broad and general way of thinking about thinking and the nature of human consciousness and language. That is, the sensitive and integrated human brain/organism has evolved to the extent that epistemological fields or states are constantly being produced and collapsing to a single state of knowledge and understanding, as interactions proceed. We look at a garden and think 'what a beautiful yellow flower', we go outside and think 'it will rain soon'. In this way, a broad description of human consciousness, language and thought emerges from our current understanding of matter, energy, society and culture.

These considerations provide an important means of thinking about formal education at all levels and how it can best be organised to support all participants (Scholl, Nichols and Burgh, 2016). For example, curriculum is not composed of slices of predetermined information that students need to accept and reproduce but rather, attempts to enable students to encounter complex and uncertain situations where answers are not known by anyone, such that field density is maximised. Creative and imaginative thinking that arises as many fields are brought to bear on a particular unfamiliar problem and then collapse or reform into one, is not penalised but celebrated in recognition of original thought based on socio-cultural experience to that time. Obviously, current approaches to assessment are totally inappropriate for this understanding. Education in its truest sense from birth to death, is accepted as a continuing process of becoming, of becoming more human, more ethical and more compassionate, of living and learning intersubject-ively with others, such that the generation and collapse of field functions is the basis of thought and the capabilities of everyone. Modern science and humanities are united in one field theory, tempered by culture, history and geography. Consciousness and language are seen as human dispositions that emerge as characteristic of growth, existing in particular regions of the universe as specific conditions for complexity accumulate.

At this point, we do not need to go into extensive detail of some of the main theorists of language, but in addition to Wittgenstein, we should note the work of Austin, Searle and Habermas. The British philosopher and linguist J. L. Austin (1911–1960) introduced the concept of 'speech act' considered to be an utterance that has a performative or action function in human communication. In this regard, Austin distinguished between a 'locutionary' act that is what we say, an 'illocutionary' act that is what we mean by saying it and a 'perlocutionary' act that is the result of what we say. This is an important distinction given the place of language in every classroom such as a statement or speech act by the teacher that 'Everyone needs to complete Chapter 9 by the end of class'. It may be that the teacher thinks this to be supportive of students and their capability of completing

a task, whereas students may infer some retribution if they do not complete. In discussing issues of language, meaning and communication, Searle (1998, p. 145) for example comments:

> When I intend to communicate, I intend to produce understanding. But understanding will consist in the grasp of my meaning. Thus, the intention to communicate is the intention that the hearer should recognise my meaning, that is understand me.

In this remark, Searle is drawing connections between speech acts and meaning and also the issue of 'intentionality' in philosophy. Intentionality is that aspect of mind whereby mental states or structures are directed at or correspond with states of the external world. Language and thought can be seen as a mode of existence that makes this internal-external connection possible. According to Searle, the object of thought does not need to exist to be represented by an intentional state, such as the characters of a fairy tale. Again, these are descriptive comments to help explain the human organism, but they are very broad and are always open to contest and criticism. At this stage however I want to propose in Figure 7.1 a schema that brings together the discussion above and which places 'language as act' at the centre of human existence, subjectivity and consciousness.

This diagram is proposed in the context of the discussion outlined in Chapters 1–6 and in particular, the influence of Hegel and Marx. For example, in his 'Critique of Hegel's Philosophy in General', Marx (1932/1959) wrote:

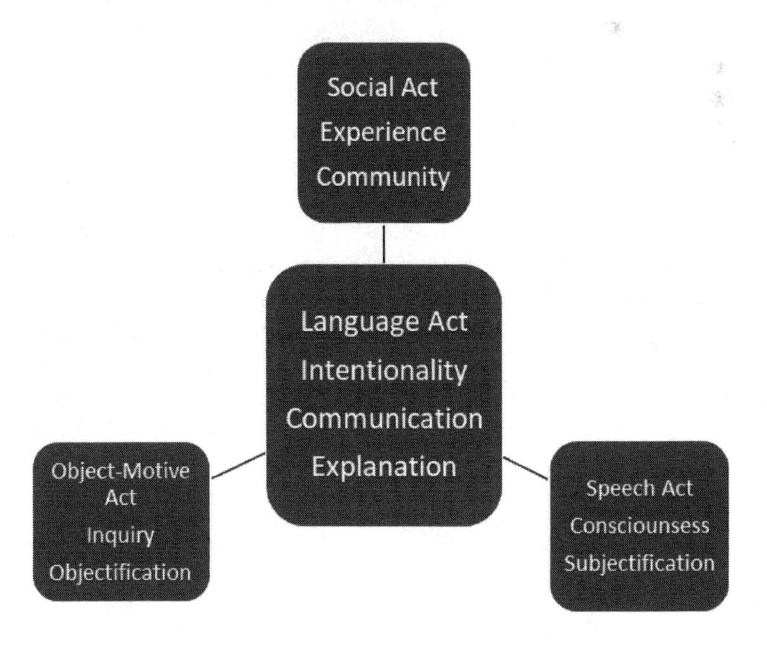

FIGURE 7.1 Connections between language act and other human acts of existence

> The outstanding achievement of Hegel's *Phänomenologie* and of its final outcome, the dialectic of negativity as the moving and generating principle, is thus first that Hegel conceives the self-creation of man as a process, conceives objectification as loss of the object, as alienation and as transcendence of this alienation; that he thus grasps the essence of *labour* and comprehends objective man – true, because real man – as the outcome of man's *own labour.* The *real, active* orientation of man to himself as a species-being, or his manifestation as a real species-being (i.e., as a human being), is only possible if he really brings out all his *species-powers* – something which in turn is only possible through the cooperative action of all of mankind, only as the result of history – and treats these powers as objects: and this, to begin with, is again only possible in the form of estrangement.

Placing the language act at the centre of our conscious humanity (Figure 7.1) relates very strongly with the comment from Marx. We conceive of language acts as emerging from the context and interrelationship of social acts, speech acts and the objects of activity, or object-motive acts. According to Kaptelinin (2005, p. 5) and drawing on the work of Vygotsky and Leontiev, 'The object of activity has a dual status; it is both a projection of human mind onto the objective world and a projection of the world onto human mind.' Further, he notes,

> The object of activity can be considered the 'ultimate reason' behind various behaviours of individuals, groups, or organizations. In other words, the object of activity can be defined as 'the sense-maker,' which gives meaning to and determines values of various entities and phenomena.

Looked at in this way, the act of object-motive does not mean objectification of the human, whose human characteristics are denied and reified, but a dialectical process of action between the objects and entities of mind and world. This process is accompanied at the same time with the act of subjectification, as necessary aspects of the dialectic, each existing in and with the other to create something new. Subjectification becomes the process of speech acts as a channel of generating internal modes or structures of conscious thought, expression and meaningfulness. Marx writes about objectification and loss of the object in a different but same understanding, as all humans can transcend the objects of domination and oppression through their labour and reach heightened awareness. Marx calls this 'objective man', where the relationship between humans, their practice and their social and physical environments are true, productive and ethical.

Creativity and creation of humanity

If humans are able to produce new ideas from their experience, then this lays the basis for creativity being a characteristic of their humanity. Knowledge does not have to be deposited from without, but is constructed from within, actively and transformingly in relationship with the cultural and physical environment. Wilson (2017, p. 3) goes further and begins his book with the statement that 'Creativity is the unique and defining trait of our species and its ultimate goal, self-understanding.' He goes on to detail creativity in this way:

> It is the ultimate quest for originality. The driving force is humanity's instinctive love of novelty – the discovery of new entities and processes, the solving of old challenges and disclosure of new ones, the aesthetic surprise of unanticipated facts and theories, the pleasure of new faces, the thrill of new worlds.

There is a strong connection with pragmatism here, with the use of words such as quest, novelty, discovery, processes, solving, challenges, disclosure, surprise, pleasure and thrill. All of these are provided in support of the notion of seeking original means of interacting with the world that is encountered in a moment-by-moment, event-by-event manner. Gardner (2006, p. 79) has a similar view, when he comments that humans are 'living creatures and living creations' where 'all human artefacts are initially created by someone'. He writes that in the view of Csikszentmihalyi (1996), creativity emerges when there is interaction between the individual who has a body of appropriate experience for the tasks at hand, the cultural domain with its customs and procedures where the task is located and the social field where comment and criticism is provided by others who are participating and interested. Gardner (p. 81) notes the definite view of Csikszentmihalyi that creativity only occurs 'when an individual or group product generated in a particular domain is recognised by the relevant field as innovative and, in turn, sooner or later, exerts a genuine, detectable influence on subsequent work in that domain'. These comments seem accurate in relation to social practice, although they are sociological and psychological rather than philosophical. I want therefore to propose my own definition of creativity that brings us back to pragmatism and the active and continuing construction of human knowledge:

> Human creativity involves the resolution of situations through dialectical emergence to enable novel and original acts of thought and action to occur and by so doing, new knowledge, new consciousness and new worlds are produced for individuals and communities. Human creativity is the basis of human creation.

In commenting on the approach of Mead to the social act, the self and the organisation of perspectives, Miller (1990) contended that 'the organism can be an object to itself and, therefore, a subject' (p. 4) and further that 'If reality is a becoming and not an unfolding of what is enfolded, then there is novelty, creativity and the emergence of new forms with their corresponding new environments with a resultant innovation of new perspectives' (p. 16). In this provocation, the human self is defined in terms of reflection, not as an entity separate to the matter and energy of the brain, but a human capability of being aware of existence, of what has gone before and of considering presenting situations for further engagement. The term perspective or hypothesis relates to scientific methodology and the argument that alignment of the entities of experience and evidence produces new entities of understanding and prospect, for ongoing investigation. The new does not emerge from a mere unfolding of what is already present, but like a chemical reaction where new compounds and by-products materialise from a bringing together of current structures, arises from rearrangements, recombinations and realignments of what already exists. The decay of elements with various half-lives into other elements, the massive release of energy from the nucleus of the atom and the conversion of a nondescript seed into a magnificent bloom, are examples of this process from nature, the process of contradiction and the unity of opposites.

When we consider creativity as a human characteristic and central to human creation, our humanity, we need to constantly remind ourselves about the nature of our own 'subject-ness' and how this can be heightened. In his discussion of art and art education, Biesta (2017, p. 15) states 'To exist as subject thus means to exist *in dialogue* with the world; it means being "in the world without occupying the centre of the world"' (Meirieu, 2007, p. 96). There are two issues at stake here, as Biesta argues. First, it is not only acting in the world that is important but understanding the right action to be taken under specific conditions for mutual interest and second, doing what is right removes oneself from the centre spotlight to a place that is right for others, for the circumstances, not only for self-interest. Biesta further notes the thoughts of Arendt (1958, p. 184) and our existence as subject 'in the two-fold sense of the word, [as] actor and sufferer'. I take this to mean that dialectical emergence does not always result in what appears to be best for each individual but may place the person at risk in relation to what is in the group or community interest. We can reflect on our discussion of ethics at this point. Consideration of risk and ethical conduct occur in Csikszentmihalyi's notion of 'flow', or the feeling of optimal experience when activity was going well in a fluid, effortless state of consciousness. He nominated nine factors that contribute to the feeling of enjoyable and purposeful experience (*ibid.*, pp. 111–113):

- clear goals throughout the process;
- immediate feedback of what is produced;
- balance between challenge and skills;
- merging of actions and awareness;
- distractions can be minimised or excluded;
- failure is not a concern;
- disappearance of self-consciousness;
- time takes on a new dimension;
- activity becomes autotelic, an end in itself.

In a moment, we shall look at music in relation to these factors. What they conjure however is a human being involved in social activity and more specifically,

SCHEMA 6: FACEBOOK POST 7 AUGUST 2017

As someone once said, 'You can take a horse to water, but you can't make it drink.' Perhaps it might also be correct to surmise that, 'You can encourage a person to think, but you can't make them learn.' Or, a little more concisely, 'It's not possible to teach anyone anything.' By this I mean that it is up to each of us to make sense of our experience rather than to have views imposed by others. This seems to me to be a fatal flaw in formal education at all levels. While it may be generally agreed that the Mona Lisa is a source of mystery and beauty, this cannot be taught. The educator may give a very expansive and expressive description of the painting, but the viewer will bring their own culture and history to bear to form their personal response. I think this is certainly true of mathematics and therefore the major weakness of school mathematics. I well remember when in about Year 10 of being fascinated by what is called 'mathematical proof', a series of steps in algebra and geometry that for example could show one angle in one part of the diagram was equal to another angle in the diagram. Not only did this give me a sense of satisfaction that I could take my current knowledge and, through a series of logical connections, show the relationship between angles (placing QED at the end, quod erat demonstrandum), but the process prompted me to think about how this was all possible, why did algebra and geometry work? After spending a lifetime in formal education, I am comfortable with the view that I have never taught anyone anything. I have stated for some time (almost from the beginning of my time in teaching) however that I have tried to establish inquiry settings that enable participants to explore important issues and ideas that they and others consider central to human knowing. I think the concept of molecular structure falls into this category and there are many experiments that can be conducted at all levels to challenge thought. How accurate such experiments are in the manner they are conducted at school and what conclusions can be drawn, is problematic, but they confront what we think, why we think that way and take our imagination in a variety of directions – provided the correct answers are not mandatory on the unit test of course and our emerging viewpoint is not rejected and failed. For me, this means that educators need to create settings that generate wonder, surprise, puzzlement and awe for everyone, so that we not only feel connected to something new, called knowledge, but there is a growing realisation that we are part of something big, bigger, something beyond our current experience that diminishes emphasis on 'self' and local and replaces it with consciousness of grandeur and the universal. Comparing angles in a geometry diagram won't necessarily do this for all Year 10 students, but it will for some, the purpose of school mathematics. For others it will be the structure of sodium chloride, the nature of greenhouse gases, the harmony of poetry, the use of the comma. Knowledge promotes wonder and awe and takes us out of ourselves, but the journey is voluntary and cannot be coerced.

social acts, that arise from conscious states, but negate conscious states, so that self-consciousness of the organism dissipates in favour of the organism as subject, capable, productive, creative. In the previous discussion, I designated this feeling or awareness as 'knowledge knowing knowledge'. Under these conditions, the creative person does not worry about failure, but is lost in the dialectic of action and thought and what that produces, to the extent that failure becomes a non-issue, the basis of further interest and investigation. The elimination of failure from schools is therefore a realistic possibility, having extreme implications for curriculum and the mass testing of students around the world. (For a general discussion of creativity in schools, see Craft and Jeffrey, 2008).

In this regard, Greenfield (2016) offers an extensive discussion of music, creativity and consciousness. As a neuroscientist, she suggests that 'music is directly comparable to language and that the uniquely human pleasure – and indeed survival value – of course must be understood best as an equal yet opposite counterpart to the spoken word' (p. 104). Her argument concerns the right hemisphere of the brain where emotion is said to reside and, linking emotion with tonality, shows connections between the tonality of music and the prosody of human speech. There is an interesting problem of dialectic contradiction that may be possible here where, according to David Huron, 'Music can never attain the unambiguous referentiality of language, nor language the absolute ambiguity of music' (cited by Greenfield *ibid.*, pp. 104–105). Greenfield raises questions regarding the harmony, rhythm, stress, intonation, anticipation and the like regarding music and language and summarises (p. 105):

> In this way, language and music are two sides to the unique coin of expression and communication that is the birthright of our species and utterly complementary in their equally important, yet different, roles. Music amplifies, exemplifies, or reinforces the course of an ongoing experience. It provides a way of being in the here and now, unlike language, which is invaluable for referring to things you cannot detect directly with the senses.

If language and music are integral aspects of human consciousness, then from the point of view of pragmatism, they both emerge from social action in response to situations and events that need to be resolved. They will both have Darwinian importance, although each will be modified in practice and form as evolution occurs. When our brain is deeply involved with language and/or music, we can be transported to other places, in thought, day dreams and night dreams that generate new perspectives on what is real. According to Mead, this is a social process where we take on the role of the 'generalised other' in making judgements and understandings of how we should act next. When we wake up after a vivid dream, or the teacher calls us back to the classroom task, it appears that the brain has taken over with a 'mind of its own' and has decided for itself what is important and what must be considered and enacted. Rather than ignore these signs, it may be appropriate to think about their significance, why they have occurred at this time and how they connect with what the organism is confronting. The neurological process by which thoughts 'pop into your head' is not known, whether or not memories are locked away in a complex cognitive filing cabinet for retrieval, or connections between assemblies of neurons form and reform and how meaning is ascribed to produce awareness, imagination, feeling and emotion. We observe what happens and create our own explanations and perspectives for the past and future.

Steven Pinker is a linguist and cognitive psychologist who sees language as an instinct, or a natural reaction by humans to the social and physical environment. In a similar manner to Minsky (societies of mind) and Greenfield (assemblies of neurons), he believes that learning occurs not through some type of 'general learning device', but by the interaction of different modules in the brain, each with their own specific purpose. Pinker (1994, p. 426) writes:

> People are flexible, not because the environment pounds or sculpts their minds, into arbitrary shapes, but because their minds contain so many different modules, each with provisions to learn in its own way.

Pinker goes on to speculate and suggest names for the types of modules and the functions they must perform. He lists fifteen candidates including intuitive mechanics and biology, number, mental maps for large territories, danger, food, individuals, self-concept, justice, kinship, mating (pp. 437–438). Significantly, Pinker argues that this model stresses commonalities among all peoples of the world. This is because for a 'sexually reproducing species', genetic variation will need to be quantitative, rather than qualitative in relation to the basic design blueprint: 'Genetic differences among people, no matter how fascinating they are to us in love, biography, personnel, gossip and politics, are of minor interest to us when we appreciate what makes minds intelligent at all' (p. 447). We now turn to a particular example of linking language, thought and creativity in educational practice.

As one of the major international testing regimes, the Progress in International Reading Literacy Study (PIRLS) is conducted (ACER, 2017). First administered in 2001 and operating on a 5-yearly cycle, PIRLS 2016 involved sixty-one countries and regions in testing and documenting the reading progress of Year 4 students. Year 4 has been chosen as the time that children in general move from learning how to read, to reading for learning. The programme aims to foster international debate regarding schooling, education policy and international competitiveness. A close examination of the approach to reading adopted by PIRLS (Table 7.1) demonstrates the understanding and application of language required of students.

In the first instance, the emphasis is clearly placed on 'literary experience' as distinct from literacy, involvement with literature and texts rather than the rules and structure of written and spoken speech. Next, subjective imagination comes into play regarding a range of social actions, connections and feelings. This is

TABLE 7.1 The PIRLS purposes of reading

Reading for literary experience	*Reading to acquire and use information*
The reader becomes involved in imagined events, settings, actions, consequences, characters, atmosphere, feelings and ideas, he or she brings an appreciation of language and knowledge of literary forms to the text. This is often accomplished through reading fiction.	The reader engages with types of texts where she or he can understand how the world is and has been, and why things work as they do. Texts take many forms, but one major distinction is between those organised chronologically and those organised nonchronologically. This area is often associated with information articles and instructional texts

facilitated by a high-level capability for 10-year-old children that combines not only 'appreciation of language and knowledge' but how these are present in different 'literary forms'. Finally, there is comment that this complex matrix of human awareness is brought about by 'reading fiction'. Thus, PIRLS clearly aims to shift focus from the conservative and behaviourist approach of reading being achieved through the acquisition of human-derived rules, or competence, to a more progressive and experiential immersion in language practice, or performance. In describing what might be expected to happen via this process, PIRLS states that students 'can understand how the world is and has been, and why things work as they do', or can note and think about bringing together the past, present and future. This could refer to a change in seasons from winter to spring, a brother or sister who becomes ill and then recovers, a new puppy that grows relatively quickly, or blowing down a pipe to make an interesting sound. Texts can highlight these processes by arranging the story or topic in a chronology of some type. PIRLS then provides detailed advice on the indicators of reading comprehension (Table 7.2).

There is often a significant difference to be observed between those who advocate a reading process based on rules and grammar and those who view language as emerging from the human process of experience and social engagement (Cambourne, 2014; Krashen 2014). While test results may show some initial improvement of children who are taught synthetic phonics, any difference with children who focus on language practice is eliminated by the end of primary school. Comprehension is a key point of difference, with the language experience group being much stronger in this area. It is significant therefore to note that in Table 7.2, PIRLS encourages a high-level approach to language at Year 4, with emphasis on correlation between information in text and information sought, inference, interpretation, idea integration and the evaluation of content, words used and the different elements of text. It is clear from this approach, that language from an early age is viewed as a creative area of human action, where activity, thought, reflection and judgement all come into contention throughout life. There is even comment on 'world view', where young children are able to consider

TABLE 7.2 The PIRLS processes of reading comprehension

Focus on and retrieve explicitly stated information	Readers are required to recognise information or ideas presented in the text, and how that information is related to the information being sought. Specific information to be retrieved is typically located in a single sentence or phrase.
Make straightforward inferences	Readers move beyond the surface of texts to fill in the 'gaps' in meaning. Proficient readers often make these kinds of inferences automatically, even though it is not stated in the text. The focus may be on the meaning of part of the text, or the more global meaning representing the whole text.
Interpreting and integrating ideas and information	Readers need to process the text beyond the phrase or sentence level. Examine and evaluate content, language, and textual elements. Readers draw on their interpretations and weigh their understanding of texts against their world view – rejecting, accepting or remaining neutral to the text's representation.
Examine and evaluate content, language, and textual elements	Readers need to draw on their knowledge of text genre and structure, as well as their understanding of language conventions. Readers may also reflect on the author's devices for conveying meaning and judge their adequacy or identify weaknesses in how the text was written.

different ideas in relation to their own feelings and perspectives. It is well known for example that children from an early age have a sense of 'fair play', or justice and what is acceptable conduct or not. What this discussion indicates then is that the educational agenda of schools within the international neoliberal context, can be framed so that human consciousness, subjectivity and awareness are recognised and that consequently, language and creativity can emerge as social acts of existence and emancipation.

Case 7. Beyond the spoken word

It was usual for the staffroom at morning recess on Monday morning to be a fairly lively place and today was no different. As Jack sat down, Bella pushed a newspaper article in front of him; 'What do you think of this,' she said waving her hands, 'that dispute about the River Estate has gone to court, who knows what will happen now.' Teachers at the school were well aware that development of a large rural block of land close by would not only change the nature of their environment, but would stop all their science, ecology and conservation programs that the students enjoyed. Let alone the annual family day and picnic. Jack scanned the article and saw that the current owner of the block was an Italian migrant who had come to Australia with his family after the war and his lawyers were claiming that his reading of English was not strong. Although it was said that he had about Year 3 reading ability and found complicated contracts difficult, he could translate

Italian into English and could actively participate in negotiations. He was saying that the contract was not what he thought had been discussed and agreed. 'I see it's our fault again,' piped up Jonathon with a laugh, 'why aren't teachers teaching children to read?' Of all topics on education and their jobs that caused both tension and discussion among staff, reading was at the top of the list and it seemed, had been forever. As a mathematics and science teacher, Jack wasn't sure why the dispute over reading had not been sorted out long before this and was unconvinced that it was possible to accurately assess levels of reading anyhow. He agreed of course that if someone was struggling, especially from immigrant and refugee background of any age, then extra help was required. He wondered about the difference between 'simple' and 'complicated' contracts and whether 'translation' meant the general idea of a paragraph, or the detailed sentences of a paragraph. It would probably be necessary to have both in a legal sense. Over the years, Jack's interest in education had shifted somewhat from his subject content area to knowledge generally and he was fascinated by how some children picked up on ideas quickly, while others, often from similar family backgrounds, appeared lost. This was particularly so in his mathematics classes. As he thoughtfully unwrapped his salad roll, Jack said to Bella, 'We shouldn't take too much notice of a newspaper report, but what do you think about the difference between legal-type words in a contract and how Angelo might speak and think?' 'Well, lawyers do use a lot of unfamiliar legalese and I suppose teachers have their own way of speaking too, but we need to be able to make sense of the language, no matter who is taking, or writing.' Jack was keen to press the point, one that he had made before, but was still trying to finalise in his mind: 'I'm sure that's right Bella,' he said quietly leaning forward for emphasis, 'but in my maths and science classes I have to somehow use new words and concepts and then go beyond them so that students can make their own meaning regardless of me; that's a first step I think. It's not only the new concepts themselves but how each student relates to them in their mind.' After a few seconds, he added, 'That's why I do experiments whenever I can and take my groups down to the block and river.' As the bell went and they moved off to their next classes, Jack was just as confused as ever. If his own thinking on these questions was unclear, how could he expect his students to read, embrace and imagine the great ideas he was trying to introduce?

8

THINKING IN SOCIAL AND TECHNOLOGICAL ENVIRONMENTS

> Physicists reconstruct grandma's narratives
> extracting identities of quantum particles
> imagining unknown worlds into existence
> splitting neutrons from their property of spin
> demanding counterintuitive understanding
> where essence is removed from social relations.

Regardless of the understandings of philosophy and science, humans live in what could be called the real world and must contend with the wide range of contradictions and experiences that are encountered. This involves life as an aspect of the universe along with all other aspects and life as a social being that has emerged on earth. Two comments or problems regarding human cognition arise from the interpretations of Chomsky and Wittgenstein. First, it is difficult to argue that instantaneous thoughts regarding the beauty of a flower occur without reference to previous cultural experience, indeed human experience that extends over thousands of years. A feeling of sadness when reading a story, or looking at a painting, does not occur in a vacuum, but within a context of human experience and activity involving nations, communities, families and individuals. It is possible to intellectualise where these feelings and emotions come from, but it is not possible to separate the various and interwoven factors of causation. In a process of idealisation, science does attempt to conduct experiments in isolation from the totality of reality, so that some indication of principles can be discerned without the complexity of social, technological and physical environments. It is also not possible to prevent humans in thinking about, discussing and investigating these outcomes either in private, quiet moments, or together in groups. In fact, when serious events happen that may cause collective delight or distress, people seek celebration or solace in each other as they express both satisfaction and support.

That is, while feelings and emotions are difficult to speak, they do not remain silent. Second, as mentioned previously in suggesting that science and metaphysics are fundamentally different, it is possible to argue that one should not dominate the other, a supposed problem that has been identified since the Enlightenment period. On the other hand, in posing this dichotomy, it separates two types of person, with two types of thinking. This stands in contrast to the concept of an integrated consciousness, capable of engaging a wide range of experience and problems, as required. I reiterate in our narrative that scientists may draw upon the extent of cultural and scientific understanding when considering their projects, especially in the design stages, in approaches to knowledge and research methodologies. The everyday work of scientists (as Kuhn termed it within the current paradigm) and technologists for example, may have close overlap with that of artisans, politicians and priests as they seek to understand the essential nature of the tasks they face and whether their purpose is defensible. This argument suggests great connections and commonalities across the universe and across societies that should be integrated across education systems as well.

In his book, *The Myth of the Machine*, Mumford (1967) discusses the convergence of science, technology, economics and political power of the time and the resulting impact on social development. He spoke of the 'megamachine' that has come about since the building of the Pyramids, the Industrial Revolution and continuing industrialisation, whereby the extensive organisation of labour, the implementation of knowledge and central control has enabled the erection of monuments to wealth and power. He described the commonly-held belief that the machine had become 'absolutely irresistible – and yet ultimately beneficial' as the myth of the machine and 'That magical spell still enthrals both the controllers and the mass victims of the megamachine today' (Miller, 1989, p. 523). This is a much more expansive view than the notion of technology being the application of knowledge for the production of goods and services and the removal of or assistance with tasks that are repetitive, dangerous to health and require manual strength. According to Hickman (1992, p. 45), Dewey's view was that 'technology can be said to be the appropriate transformation of a problematic situation, undertaken by means of the instrumentalities of inquiry, whatever form those instrumentalities may take'. Hickman goes on to explain:

> To put this another way, inquiry is a technological activity because where inquiry takes place there is a shift from passive acquiescence toward the beginnings and endings of nature, its contingences, to the active construction of artefacts to effect their control. Immediate use and enjoyment give way to the production of consequences.

Dewey used the term 'instrumentalism' not only because of the semantic difficulty of the word 'pragmatism', but to show the use of tools such as language, thought, microscope and paint brush in all human activity and that artefacts arising both intellectual and physical from such activity are constructed, novel and open to interrogation. We thus have a view of technology as a process of transformation and production when the human organism confronts a problematic, indeterminate situation from the natural or real world that needs to be changed or become determinate through application of cognitive and physical tools. There is a dialectic between what is known or what we think we know and what is transcendent and unknown. For educators, this understanding and practice of technology as an epistemological construct is important whenever a new topic is introduced.

Seymour Papert (1928–2016) worked with Piaget in Geneva 1958–1963 and was considered by some to be Piaget's successor. Papert was a mathematician and became interested in the connections between children's learning and computing. At this stage, computers still consisted of large consoles, computing capacity was limited and data entry occurred through punched tape or cards. Being guided by Piaget's theories regarding human structures of thinking and learning, experiential learning and child-centred learning, Papert set about designing a computer environment that enabled greater participation by the child and more control of what the computer could accomplish. By this time, smaller, desk-top computers with keyboards were becoming available and outputs could be graphical and screen-based, rather than paper. Papert's research and work resulted in development of the Logo language, a robotic 'turtle' that moved around on the floor and in association with the Palo Alto Research Laboratory in California, a device called a 'mouse' that could also 'point and click' on the screen to control what the computer was doing. Through the use of a pen on the bottom of the turtle robot, or on the screen, children were able to experiment with their own ideas by changing the length and direction the turtle moved and, by so doing, see their 'ideas in action'. Papert called this approach learning through 'mind-sized bites'. There is no doubt that the Logo philosophy was revolutionary and quickly gained the attention of many children and teachers around the world. It enabled children to see their ideas come alive on the screen and the effect of the changes they made. Unfortunately, Logo often found itself immersed in the conservative boundaries of traditional education and the possibility of open-ended experiments for children was restricted. Papert then became associated with the Lego company and the first edition Lego Logo robotic kits was manufactured. Again, these were adopted by teacher enthusiasts, but the advent of new generation robotics into schools had to wait for many years until quite recently for their recognition.

In his recent book mentioned previously, Resnik (2017) makes a strong case for what he calls 'Lifelong Kindergarten'. Similar to the Action Research Spiral (Kemmis and McTaggart, 1988), he proposes the 'Creative Learning Spiral' that he has seen repeated over and over again by children in kindergarten: imagine, create, play, share, reflect, imagine. In a wonderful first paragraph to his book, Resnik (p. 1) in effect summaries what we have been discussing in previous chapters and bears citing in full:

> Human creativity moves hand-in-hand with technology. Over the course of history our tools have evolved from flint knives to the Large Hadron

> Collider and beyond. Whether they are mechanical or digital, simple or complex, tools facilitate our creativity in two ways. Initially, they extend our bodies and enable us to do things that are otherwise difficult or impossible. With ploughs we can cultivate, with telescopes we can see, with engines we can travel far beyond the limits of our unassisted bodies. But tools do more than extend our bodies: they expand our minds. Technology facilitates ideas that might otherwise be inconceivable.

We can see here the influence and legacy of Dewey, Mead, Piaget and Papert as they have struggled to explain and theorise human existence and how we have become what we are. We see the connections between creativity, technology, tools and minds. Papert pushed these linkages further when we asked why schools are concerned with pedagogy, the art of teaching, but not the art of learning. He suggested a new word, mathetics, from the Greek 'to learn' and made the following distinction: 'When I learned French I acquired linguistic knowledge about the language, cultural knowledge about the people and mathetic knowledge about learning' (Papert, 1995, p. 10). He went on and introduced another Greek notion of 'heuristic', that is to use a 'rule of thumb' where approximations or imprecise methods based on practical experience are used to make progress on resolving a situation. Supporting the work of Polya in problem solving, Papert discussed the idea of taking your time to think through problems, considering similar but different situations and breaking the problem down into manageable pieces, of playing with ideas. These types of concepts are central to the design of the Logo programme and the Lego Logo robotic kits. What Papert was attempting to do I think was to provide an overall concept for an open, experiential approach towards learning in schools – mathetics – supported by a way of considering problems of all types – heuristics. This difference can be illustrated by how we distinguish between creativity and creative thinking in schools. Given our definition of creativity from the previous chapter, we can relate human creativity to mathetics regarding the imaginative and innovative production of objects and artefacts in the working through of events and situations. Creative thinking on the other hand can be thought of in heuristic terms, or personal intellectual guidelines, such as that defined by the Australian Curriculum (ACARA, 2017): 'Creative thinking involves students learning to generate and apply new ideas in specific contexts, seeing existing situations in a new way, identifying alternative explanations, and seeing or making new links that generate a positive outcome.' It will be a difficult task to establish school mathematics for example in a mathetics and heuristic manner and will require that all set knowledge in the curriculum should be opened up to question and experimentation. It has been mentioned previously that current knowledge is respected and appreciated, but it needs to be subject to thought and critique by all participants as they proceed to engage their environments with their own thinking and practice.

If notions of mathetics and heuristics are important for learning, then they are important for the role technology and computers have in learning. Papert suggested the concepts of computer 'microworlds' and 'hyperworlds' to assist with this understanding. A microworld is conceived as a 'medium for making simple, restricted worlds' where 'learning can take place without being hampered by the complexities of the real world' (Papert, 1996, p. 56). He wrote (p. 59):

> The guiding image of the microworld is a 'world' limited enough to be thoroughly explored and completely understood. It is the right kind of place to learn to use knowledge that requires deep mastery. In an analogy between ideas and people, microworlds are the worlds of people we know intimately and well.

In contrast, hyperworlds are large worlds of 'loose connections' between idea acquaintances rather than ideas as close friends. Using the Internet to quickly gather information, to check facts and to gain an overall view of an issue rather than to seek more substantial knowledge comprehension shows a hyperworld in action, surfing the net as it is called. Wikipedia and Google are examples of such application where users can constantly make quick references to monitor their progress. Social media is a general case in point where the difference between computer microworlds and hyperworlds is stark. Platforms such as Facebook, Twitter, Instagram and the like are used by many millions of children and adults around the world and constitute a marked change in human communication, but in general, appear to be superficial in educational outcome. There have been examples where social media has been employed to break news of significant events including war, natural disaster and political occurrences, instant communication that would have taken weeks or months in the past, if at all. In terms of learning however it seems that social media at best falls within the category of hyperworld, making information available to readers if they are interested. Conversely, it may be that there are few examples of computer microworlds as envisaged by Papert, mathetic and heuristic environments that draw humans into intimate knowledge in new ways, such that they become aware of their personal and community learning and their relationship to other objects of existence. What the computer can do, or at least, what educators should demand that computer power should be able to do in the hands of children at school and at home, is to enable the 'social act' of Mead to be implemented and in addition, to facilitate the view of Piaget that abstract ideas need to be made concrete for children. Concepts of 'act' and 'concrete' are not easy to understand and then encourage in the somewhat fabricated settings of schools but, like democracy, they can be supported and approached (Lawy *et al.*, 2010). For his part, Papert suggested placing ourselves in the shoes of the child to get a better grip on what is required: listen carefully to what the child is saying when trying to untangle how an equation works, observe what first steps are taken when writing a poem, what features are incorporated into a robotic design, how are decisions made when experimenting with a basketball? These discoveries may uncover prospects for microworld construction for all children.

Artificial intelligence and education

Expert systems have been around for a long time. These days, the expression 'algorithmic decision making' might be used to indicate computer programs that can bring together a number of features involving a particular issue and then provide comment or advice on how the human operator could proceed. Programs that underpin Google, Amazon and the Internet itself can be easily understood in this way. When a reader asks a person behind the counter of a bookshop for information regarding the latest crime novels, the mathematics that shape Amazon can do almost exactly the same based on trends, new publications and sales. What the computer cannot do (except in very perfunctory terms) is comment sagely on the quality of the writing, the emotion generated and comparisons with previous books by the same author, or different authors for each reader: 'What is it that you like about her writing?' A very similar process is available in the medical field. Doctors have long been able to consult an extensive data base by logging symptoms and receiving a print-out of possible causes. The argument here is that all doctors (and all professionals) may not be completely up to date with the latest research, treatments and diagnosis from around the world, whereas an electronic data base can be constantly up-dated for immediate perusal. In terms of business applications, programs are available that allow for 'algorithmic trading' (algo trading) whereby a computer is able to automatically authorise small aspects of product orders, or orders based on the variables of time, availability and volume. Computer programs are also available to guide large investment decisions, although the importance of ensuing responsibility still rests with the human may be in doubt. Recently, the Australian Securities and Investments Commission (ASIC, 2017) announced that it wanted 'financial services firms deploying algorithms to make sure their logic is explicable to customers and regulators and that a human being is made responsible for any problems that might emerge with the computer code'. There are two implications arising from these examples. First, artificial intelligence programs are finding their way across all sectors of society and there is little to suggest that this will decrease. Second, the need for human understanding and control of such systems will remain imperative. Despite what might be called 'low level' application as above, the use of artificial intelligence for military and industrial application will drive technological capability forward and there will be inevitable spill over into the social domain, including education.

As we have discussed previously, artificial intelligence is concerned with human-like behaviour being generated by computer-based technologies. Famously, the mathematician Alan Turing suggested that when a human cannot tell the difference between machine output and human output, then the machine is intelligent – intelligent, but not human. That is well and good, provided that we have an ethical, shared and accurate understanding of what constitutes human intelligence and human-like behaviour. This may appear to be obvious when on the assembly line, piloting a fighter plane, or sending a reminder note to pay an account, but acting with human consciousness, subjectivity, compassion and fairness is another

SCHEMA 7: FACEBOOK POST 2 DECEMBER 2017

We have always had a choice about how we embrace technology: the violin, hammer, paper and pencil, spark plug on the one hand, bombs, bullets, mustard gas, security cameras on the other. I have noted before how many years ago I raised the future of artificial intelligence – simulated life with human-like outcomes – long before Stephen Hawking and whether we should decide as a society not to proceed in that direction – some possible perhaps likely outcomes were too obscene to contemplate. However the military assured that was not an option with intelligent drones deciding between friend and enemy and where to drop their payload being the latest AI manifestation. I think we now have the chance to decide what we are going to do about augmented reality (AR) – changes to daily reality through computer application – for the same reasons: where might it head ultimately? There are a number of broad areas for consideration I think:

While it may be possible to think of situations that are mildly augmented for a reasonable purpose, this may not always be the case. That is, do we have a philosophical overview and a defensible position on human impact? Computer games may be a case in point, where violence and aggression can dominate in comparison with other human traits such as kindness and co-operation (how exciting is that on a publicity video?). Then there is the question of pornography on the Internet and its availability to all. Opinion and research varies on this point that is not really resolved by a censor-type system of categorisation of computer games and applications similar to film. Any attempt at prohibition or regulating the AR field will inevitably lead to a black market. It is of course entirely feasible to develop a set of criteria about an acceptable, human-centred approach to AR, but with only prospects of mild or superficial success at implementation.

I am not aware that we have a developed set of ethical criteria regarding the violin, hammer, paper and pencil and spark plug let alone bombs, bullets, mustard gas and security cameras (we could discuss the role of the UN here). We certainly do not have any criteria let alone public debate about the use of drones for war and aggression. It is now intended that small drones will be part of the kit of infantry, able to drop explosives over the horizon. I suspect that drones (including AI drones) of varying size and capability will be purchased by the Australian military without any public discussion whatsoever. We can attempt to raise issues regarding the approach to knowledge that AR might embody, how it will deal with culture, multicultural and cultural stereotyping, the inclusion of Indigenous perspectives from around the world, ensuring social class, feminist, location and disability considerations, recognition of and respect for different points of view rather than dismissal, understanding the creativity of children and many more. That is, AR should enable the complexity and diversity of human society to be acknowledged and not be denied or simplified.

From my viewpoint, it seems that the technology available for schools and families today does not enable AI principles to be implemented in markedly useful ways. On the other hand, I imagine whatever computer power is necessary for military purposes and space exploration is made available. It may be therefore that in this initial phase of AR public application will suffer a similar fate, that is what we would like to be able to do for educational purposes for example, will not be possible. It will be a watered-down version but marketed vigorously as usual. In fact, we are still unsure of the educational impact of computer-based applications for all children, despite some enhancement at some levels. So, I ask the profession, AI, AR, VR are here, in our classrooms, or are they? Neil

matter. Robinson (2017, p. 187) for example, comments that a culture of innovation within human organisations involves a relationship between three processes:

- imagination, or the ability to bring to mind events and ideas that are not present to our senses;
- creativity, or having original ideas that have value;
- innovation, or putting original ideas into practice.

These are familiar human actions from the point of view of pragmatism and human inquiry, but may be difficult to program into a computer, let alone to consider in relation to each new situation as it is encountered. Deciding whether to enable a missile to make its own decisions about targets and the enemy once fired from location thousands of kilometres away may have military logic, but human stupidity as a consequence. In raising the question of what it means to be human in the age of artificial intelligence, Tegmark (2017, p. 81) agrees that 'Today's artificial intelligence tends to be *narrow*, with each system able to accomplish only very specific goals, while human intelligence is remarkably *broad*'. While Tegmark says that we cannot predict the future, he does accept that the trend towards 'superintelligence', or technological intelligence that will surpass that of humans, will continue. It is not only science fiction writers and film producers who describe a world inhabited by cyborgs of superintelligence interacting with organism of human intelligence, the replacement of body parts by technological substitutes and ultimately, the uploading of human consciousness into robotic forms. The problem with this latter projection of course as we have discussed is that we have spent many centuries trying to work out the nature of human consciousness and to upload something we cannot identify is a trite difficult. Can we upload memory without knowing how memory exists? Remember that

Weizenbaum alerted us in 1984 to the 'disgusting' prospect of transplanting human heads and whether or not technology should be permitted to move in this direction. Tegmark however offers the challenge that engineers need not necessarily follow the road evolution has travelled to human intelligence but may discover simpler solutions. These solutions do not need to be understood. I reported previously the capacity of the computer program AlphaZero Go to become an expert Go player, by the random accumulation of experience in playing the game. What this means is that we can build machines that exhibit human imagination, creativity and innovation without knowing how they work, or indeed, without fully understanding their character and meaning in the first place. In his discussion of the need for a 'democratic awakening' in the age of artificial intelligence, Boyte (2017, p. 37) notes 'a yearning for human relationships as an antidote to increasing impersonality'. It is entirely possible that intelligent robots with approximate empathetic, cultural and political circuitry can be released and work alongside their human counterparts in very useful ways, but as part of their remit, to take on a 'life' of their own, be able to learn about society from their experience and either 'agree' with current social and ethical practice, or not. As was commented by the sentient computer HAL in the film '2001: A Space Odyssey', when asked to explain a number of malfunctions of the space craft, HAL replies 'human error'.

Sociality in the technological age

Pragmatism will not be able to clearly separate sociality and technology, given the definitions above and the understanding of technology in the early part of the twentieth century. Mead's notion of sociality as a form of existence between the past and the present contains a phase of adjustment between the old and the new giving rise to the emergent object. He describes this difficult concept as follows (Mead, 1938/1972, p. 654):

> Man sets the universe out there as like himself, identical in matter and substance. In considering the observational field, we get characters of the object like ourselves. This shows the nature of the inside of the object. In a certain sense, one deals with an outside, but, when one gets hold of it, one has a completely congruous experience – one puts into it the attitude which helps to get at it in terms of one's self. This in short reveals the social nature of consciousness and the fact that the reflective process itself employs a mechanism of social conduct.

If we take one example of object as stone, then turning over this object in our hand enables us to view superficial appearance, or outer and to conceptualise material substance, or inner. In so doing, we bring to bear the sum total of our experience involving cultural background, community stories, personal interest and the like on imagining the inner composition of the object. This is somewhat of a mystery, while the outer is more familiar involving shape, texture, colour, rigidity and weight. We think about this object dialectically, our thoughts connecting with the object and the object connecting with our thoughts as we experience its effects. In this way, we can think of being inside and outside the object at the same time. A second example could involve the object of language. As we experience the effects and properties of the outer surface, that is sound, utterance, tone, loudness, we also feel being drawn inside language with its evocation of thought, expression, emotion and meaning, impacting on our connection with and understanding of language at the same time. In considering language as being a mode of consciousness, the experience of acting with the objects of existence, or more correctly social existence, we develop human subjectivity, an awareness of being subjects of our own existence. Living in this way makes the world more knowable; we become closer to the trees and rivers, the beauty of a child's question, a loved-one's soft word of gratitude. To illustrate this point, a specific aspect of language known as poetry produces strong feelings and emotions within the human organism, to the extent that we are transported into different worlds, or systems. A well-known poem originating during the 1880s describes an Australian office worker in a crowded, impersonal big city dreaming of living in the bush, freed from the stultifying constrictions of the office, embracing the great Australian landscape (Paterson, 2017). It says in part:

> And the bush hath friends to meet him, and their kindly voices greet him
> In the murmur of the breezes and river on its bars,
> And he sees the vision splendid of the sunlit plains extended,
> And at night the wondrous glory of the everlasting stars.

Known by most Australians, the poem arouses images of a stockman on horseback riding near the river and its sandy bars, camping on the river bank at night and having a brilliant canopy of stars overhead. These lines are some of the most recognised in Australian literature and show optimism for a satisfying future, the 'vision splendid'. Feelings of democracy and peace that arise in the collective breast from a reading of this poem demonstrate Mead's concept of being 'inside' the object of experience, expressed as language and being able to utilise a gathering of words as the 'outer' surface of that experience. It is possible and in many cases most likely, that humans will not come into an explicit and intimate relationship with the objects of experience in the way just described. This may be for political reasons, for social circumstances, or because of institutional and family restrictions, but the general process of acting with the world will establish the basis of human connection with the surrounding environments and with each other. These philosophical issues raise many questions for the educator and the design of teaching programs but are often ignored.

A recent international phenomenon in education concerns a grouping of subjects known as STEM, that is science, technology, engineering and mathematics. Originating in the United States as a means of strengthening mathematics and science education in particular for economic competitiveness and increased workforce skills, the popularity of STEM among policy makers and educators has spread world-wide. A national STEM report in Australia for example, notes that 'schooling should support the development of skills in cross-disciplinary, critical and creative thinking, problem solving and digital technologies, which are essential in all 21st century occupations' (AEC, 2015, p. 3). It is strange indeed why these subjects have been isolated, although they were initially identified as METS in the United States before the advent of digital programs in schools. Various combinations of school subjects have been attempted in the past but usually for educational rather than economic reasons. For example, the notion of general science or general mathematics brings together the different topics of each such as physics, biology, chemistry for the former and algebra, trigonometry, geometry for the latter. This provides an integrated approach to knowledge where concepts and ideas can be bridged and support each other. Similar attempts can be seen under the heading of humanities, languages, arts and the like. While STEM on the other hand appears to be a loose grouping of subjects that have little basis for integration based on their selection for employability and other political reasons, there is debate regarding a more integrated approach. In an obvious misunderstanding of the original purpose of STEM, educators have added other subjects to broaden perspective such as STEAM where the A designates Arts and METALS, where the L stands for Languages. It is unlikely that the STEM movement will take into account our discussion above, where the sociality of human experience is explicit, emphasised and understood. We could take any example here such as artificial intelligence, virtual reality, or 3D printing, but let us consider solar energy. Presumably, STEM projects will include solar energy at some point, but will they encourage staff and students to look beyond the surface to inside conceptual understanding? Notions of energy, light, electron, photon and heat are complicated and abstract, but essential for this topic. They amount to a detailed philosophical study based on a range of scientific knowledge that may or may not be accessible to young learners. Any science teacher is in the same position, but STEM teachers will need to show how they will approach this question drawing on the integrated knowledge of each subject area. If not, then the basis of STEM lies in tatters. This criticism may sound harsh, but the popularity of STEM world-wide needs to be justified in epistemological terms, or it will remain another ill-informed, non-research-based educational fad.

At this stage of our narrative journey, we are in a position to summarise some of the key ideas that we have tentatively drafted and where we might head in Part III of the book. Figure 8.1 indicates that the concept of dialectical emergence is supported by four broad areas of human practice and thought, those of earth-sky, inner-outer, backwards-forwards and alignment-realignment. These will be briefly considered in turn and extended in Part III.

Discussion of these concepts remains essentially descriptive, but they provide a means of thinking about our own thinking both personal and community and how various social practices influence what we think and how we act. In writing about the significance of Papert's work, Turkle (2017) comments:

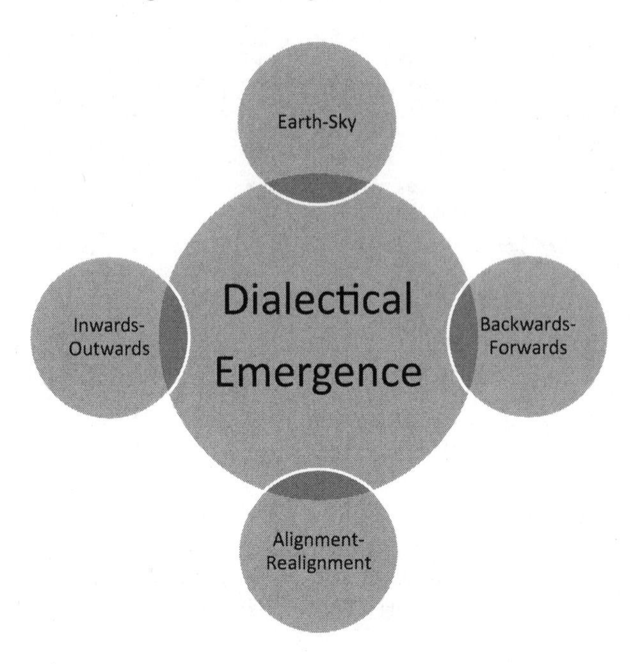

FIGURE 8.1 Dialectical emergence and supportive practices

> We love the objects we think with; we think with the objects we love. So teach people with the objects they are in love with. And if you are a teacher, measure your success by whether your students are falling in love with their objects. Because if they are, the way they think about themselves will also be changing.

Such objects can be hypostatised or not. For the purposes of discussion in this book, we have taken dialectics to mean a general view of motion and transformation of social and physical objects and processes in the universe and contradiction to be the working out of specific relationships involving processes such as the unity of opposites and negation of the negation. Contradictions are envisaged as existing in reality and that the most appropriate way to understand the movement of that reality is to study the development of those contradictions. As noted by Eagleton (2016, p. 7), 'From ants to asteroids, the world is a dynamic complex of interlocking forces in which all phenomena are interrelated, nothing stays still, quantity converts into quality. . . . and reality evolves through the unity of conflicting powers.' The most extreme notion of dialectic and contradiction concerns the increasing expansion of the universe itself and how, ultimately, the strength of expansion will overcome nuclear forces holding atoms together, resulting in the universe being ripped apart, progression from the 'Big Bang' to the 'Big Rip'. In this context then and guided by Dewey and Mead:

- Backwards-forwards describes how we connect past social experience with current encounters so that a pathway forward or resolving issues and contradictions can be postulated and constructed.
- Inwards-outwards specifies the contradiction between what we think we know arising from inner consciousness, subjectivity and language in contrast with our engagement with the natural or external worlds, what we don't know at each moment.
- Earth-sky symbolises the groundedness or rootedness of our being in social practice that then conceptualises and enables human aspiration or melioristic action to make a better world, to progress beyond natural and imposed change.
- Alignment-realignment refers to the constant movement and adjustment of thought and action as the resolution of contradiction works itself out on personal and community levels, bringing forth the new and novel for mutual and public benefit.

In this way, human creation is an integrated materialist process of dialectical emergence over time within which, every citizen of every culture and background is connected, situated and actively engaged.

Case 8. Under water adventure

It looked like it might be a reasonable day, not quite as hot as yesterday, although that warm north wind could make things unpleasant. Colin was usually up early during the summer school holidays and couldn't wait to get down to the beach. He found it difficult to go for a run in the mornings – getting into gear he called it – and waited until later in the day, perhaps the evenings. He was saving up for a surf board and had a way to go, but until then, loved body surfing and the sensation of the ocean as its waves powered him towards shore. Colin was somewhat of a loner who enjoyed his own company, so he had not joined the local surf life-saving club, much to the disappointment of his father. It was enough to be outdoors no matter what the weather, to run, to explore, or merely walk and observe the changing nature of the ocean. An occasional seal would come into the bay and he felt he knew the stingray that wandered around the breakwater when fish were being cleaned. What did strike a chord with Colin was the advent of a small number of scuba divers who would appear during the warmer weather. He would watch them getting prepared and strapping on their various items of equipment before swimming out to the reef and disappearing. Even in summer, the water along the west coast was moderately cold making a wet suit essential, something that was far too expensive for Colin and his family. But as he sat on the rocks, he often wondered what it would be like to dive beneath the surface and to explore what was there, like one or two films that he had seen. No-one that he knew went scuba diving, so he had to rely on his fertile imagination and Jules Verne and Jacques Cousteau for advice. Then, one day, out of the blue,

Colin had an idea. He was passing by the local sports shop when he saw a plastic snorkel in a corner of the window and, on the spot, decided to take a risk. It was a simple device, a curved tube with one end to go in your mouth and a ping pong ball arrangement at the other that would stop water getting in when under the surface. He had to purchase a face mask and flippers as well that would drain his savings, but, what the heck. Colin felt the north wind on his back as he walked into the waves and adjusted his mask and snorkel, thoughts of World War II espionage in his mind. He had decided to swim near the small sandstone island that he often visited and had heard about deep holes and currents that could be dangerous. With a little apprehension, he put his head underwater, breathing through the snorkel and gazing through the face mask. He was amazed. This was a world he had not realised existed, even though he lived nearby, the greens and blues of the water crystal clear, yellow sand and rocks, brown clumps of seaweed gesturing and swaying, small fish looking at him with disdain. He became aware of the sound, both silence and slight as he moved about, even attempting a small dive to check that the ping pong ball worked. As Colin walked back to the beach through the shallows, he took off his mask and snorkel, excited at what this small plastic tube had just enabled him to do, but also frustrated that it had taken him so long to look where before he could not see. He wondered what it must be like out at the reef and to be able to spend minutes or even hours beneath the waves he knew so well. Scuba gear was still beyond his price range and the experience of his family, but you never know, they might be gripped by the idea as well.

Knowledge Exemplar 2

Chapters 5–8 raise many difficult questions about the 'inner and outer' of our humanity that philosophy has not resolved to the satisfaction of all. I hope that the schema and case have provided some avenues into my thinking about these issues. What I have attempted to do with Exemplar 2 in Table 8.1 therefore is merely suggest some indicators of the objects of contradiction that we may observe in social practice. In reading through the indicators, one is struck by their interconnectedness and their emergence from the acts of daily experience whether at school, at home, or participating with others in community organisations. They provide a broad framework for understanding how we relate, live, work and learn as human beings. As mentioned throughout the book thus far, it is difficult to appreciate our existence and our humanity from introspection for, according to Habermas (cited in Zizek, 2014, p. 92):

> The attempt to study first-person objective experience from the third-person objectifying viewpoint, involves the theorist in a performative contradiction, since objectification presupposes participation in an intersubjectively instituted system of linguistic practices whose normative valence conditions the scientist's cognitive activity.

TABLE 8.1 Knowledge Exemplar 2: contradiction

Objects of Contradiction	*Indicator 1* *Community*	*Indicator 2* *Family*	*Indicator 3* *Personal*	*Indicator 4* *Country*
Subjectivity	Connection with events, debates, proposals	Knowledge of relatives, stories, histories	Feelings of delight, excitement, sadness, frustration	Experiencing outdoors, animals, plants, water, people
Consciousness	Being aware of knowing location, happenings	Closeness from sensations of care, respect, responsibility	Immersion in activity, knowing that, knowing how	Being part of holistic environment
Creativity	Acting together for mutual benefit	Recognition of changing tensions, understandings, comfort	Taking initiatives to resolve issues for progress	Interactions with all components for sustainable outcomes
Language	Construction of meaning as basis of general acting with others	Productive act to establish relationship with intimate group	Connecting with direct experience for understanding	Enabling expression of thought regarding environment
Technology	Artefacts arising from mutual experience	Thinking about activities involving close associates	Feelings generated from use of artefacts, entities	Connecting with the land via thinking, objects, apparatus
Sociality	Comprehending from different perspectives at once	Emergence of new objects for group identification	Adjusting action in relation to external considerations	Awareness of linkages between social and physical objects
Education	Lifelong experience of social practice	Specific actions informed by social class, race, ethnicity, gender	Alignment of perspectives in relation to what is known, what is not	Understanding of transformation of matter, energy throughout universe

Habermas is pointing out the difficulty of attempting to explain our own understanding from the standpoint of our own understanding and using words that do not conceptualise at deeper levels of cognition – trapped in effect by our own understanding and consciousness. However we will attempt to encounter more detailed description of such human activity in Part III, specifically regarding the practice of formal and informal perspectives and education.

PART III

Looking to the earth, looking to the sky

Having structured our narrative around looking backwards and forwards, or taking some key educational ideas and connecting them to ongoing activity and looking inwards and outwards, or attempting to understand how human subjectivities influence our thinking and practice, we now consider how these perspectives are enacted in distinctive social spheres. The notion of 'looking to the earth' is conceived as establishing or grounding social practice and philosophy that generates awareness of oneself and current circumstances of personal and community experience. 'Looking to the sky' enables generation of aspiration, challenge and purpose based on personal understanding and encountering new circumstances for exploration and experiment. Part III begins this discussion from the point of view of Indigenous peoples, where there is a sense of Indigeneity or characteristic socio-cultural perspectives that constitute awareness of what it means to be Indigenous. Next, issues are raised regarding feminist understandings of the world and whether or not it is possible to identify feminist knowledges that will then impact on educational practice. In considering the production of knowledge itself, the concepts of formal and informal approaches to research are noted together with methodologies that are more creative, imaginative and open than traditional processes. Finally and in theorising the discussion and conjectures canvassed from throughout the narrative, Chapter 12 proposes a general epistemological framework for guiding engagement with the social and physical worlds. Education is envisaged as a philosophy of practice occurring in daily life for all citizens of all cultures, as well as in systematic professional projects and studies. This framework has historical and democratic significance in that it builds upon the philosophy of pragmatism but moves beyond in explicitly enabling the participation of different social groupings that are often excluded from mainstream activity, encouraging the respectful incorporation of community and cultural experience in all forms of knowledge production. Implications for the reorganisation of schooling are penetrating at the systemic and local levels, ultimately impacting on the social stratification of education with democratic and humanistic revolutionary consequences.

9

INDIGENOUS PHILOSOPHY AND KNOWLEDGE

> Comprehension of the essential unity
> is obstinate refusing to die
> not to eliminate difference but to
> enhance appreciation and courage
> in confronting the fearful and unknown
> aspirations collide to chart a common future.

Ensuring that all families are included in democratic, communicative and participatory formal education remains one of the most difficult social questions around the world. Different groups live, understand and realise differently, groups that may be large in number such as working class people, or few in number such as Indigenous peoples in first and second world countries. If however close working relationships are established between marginalised and dominant groups, then systems of schooling and teaching can be altered to respect diversity and plurality (Bishop, Ladwig and Berryman, 2014). Changing policy and practice does not occur easily and requires committed and painstaking effort over long periods of time. For example, the notion of Indigeneity is difficult to grasp for non-Indigenous people and may be interpreted differently by different Indigenous communities. It can be taken to include questions of origin, ideas regarding connections with the land, cyclical rather than linear time, the centrality of family and community rather than individuals, the role of Elders and other community members as knowledge holders and story tellers and the practices of law and lore. These notions constitute a different world view to dominant neoliberal concerns and explain why many Indigenous children find formal schooling alienating. Conversely, they provide avenues to meaning that non-Indigenous children may appreciate and adopt in full or in part as they try to come to grips with the complexities of school knowledge. This chapter argues that a rigid curriculum and approach to learning about ourselves and our place in the universe is inherently undemocratic and inequitable, whereas a more flexible framework that allows for diverse viewpoints in formation can be self-regulating as it seeks to respect the knowledges of all children from all backgrounds. Creative, action-oriented thinking about Indigenous education and its place in the intended curriculum makes the enacted curriculum more responsive to the needs of all groups of children and ultimately provides the opportunity for evolving curriculum of a new type.

For Indigenous peoples, we take the notion of 'looking to the earth, looking to the sky' as an awareness and perspectives of Indigeneity, of what it means to be Indigenous. A basis for 'thinking beyond' the immediate becomes possible. This is not the same as 'what it means to be Spanish' for example, or 'what it means to be female', but results from a combination of history, culture, community and relationship with the land that is fundamentally different to a non-Indigenous outlook. Given this framework and in terms of knowledge, the comment from Yunkaporta (2009, p.2) should be noted:

> In our world the deepest knowledge is not in words. It is in the meaning behind the words, in the spaces between them, in gestures or looks, in meaningful silences, in the work of hands, in learning from journeys, in quiet reflection, in Dreaming. The eight ways were tested on journeys following the river along a codfish song line linking to the Murray River, tested in ceremony, tested in the carving and use of tools to represent them. This silent knowledge was explored with the hands and the feet. A lot of this knowledge can't be shown with words in a book like this—but in our way it would be up to the Aboriginal listener (in this case reader) to fill in those gaps themselves, to fill it with their own cultural knowledge and teaching experience.

Indigenous philosophy is generally described as involving the interrelatedness of all aspects of the universe, cycles of experience, 'seeing' the perceptions of nature, intimate connection with and belonging to the land, family and community stories and oral traditions, careful listening and patience and the knowing of relationships between events. Traditions of living are passed on from generation to generation by Elders and community members. They are clearly distinct in the main from those of non-Indigenous society and regular schooling that rely on linear cause and effect, detailed analysis of specific and isolated issues to relieve doubt, and a distinct separation of the physical and metaphysical. For these reasons, the process of colonisation including the forcible removal from land placed considerable pressure on tribal knowledge, language, and customs and remains a major cultural disjunction for Indigenous peoples today whether living in urban, regional, or remote communities. The dominant society will always dominate making it extremely frustrating and often impossible for marginalised groups to participate fully in social life while at the same time remaining true to community beliefs and practices. For these reasons, a democratic and equitable society must establish ways of recognising and respecting Indigenous history, language and customs in all appropriate social institutions and procedures to provide cultural identification and sustainability.

In considering the above principles, the work of two Aboriginal and Torres Strait Islander Australian scholars has been drawn upon in constructing a research framework. First, Martin Nakata has developed the concept of the cultural interface where he describes this 'contested space between two knowledge systems' (Nakata, 2007, p. 9) as being not clearly Aboriginal and Torres Strait Islander or non-Aboriginal and Torres Strait Islander. This could be described as a 'liminal' (Turner, 1967) consciousness as understandings become more variable and are challenged and questioned by changing circumstances. Nakata suggests that the Aboriginal and Torres Strait Islander epistemological constructs of knowledge are embedded in land and place for many Aboriginal and Torres Strait Islander people, as well as 'ways of story-telling, of memory-making, in narrative, art and performance, in cultural and social practices' (2007, p. 10) such that there is a constant process of acting and transforming for family and community concerns. Second, in discussing '8 Aboriginal Ways of Learning', Yunkaporta and Kirby (2011) indicate that at the core of Aboriginal philosophy is 'bringing culture into the "*how*" not just the *what* . . . We're learning *through* culture, not just *about* culture' (p. 206, italics in original). This is recognition that culture is central to learning and simply cannot be added on to the development and implementation of assumed Aboriginal perspectives regarding learning and schooling. In combining understandings of the 'cultural interface' of Nakata and the '8 ways' of Yunkaporta, our thinking attempts to locate its purpose and methodology within a broad framework of Aboriginal and Torres Strait Islander epistemologies that engages dominant knowledge and pedagogies.

Holistic approaches to learning that are infused with culture, language, and community intersect closely with the pragmatic philosophy of John Dewey discussed previously (Dewey, 1910/1977). To review, in relation to Indigenous sociality, learning arises from the continuing experience of living and thinking carefully about the effects of experience. That is, knowledge emerges from the resolution of disrupted habits and customs by the adjustment or alteration of belief, or the stabilisation and formation of conventions in response to the resolution of doubt. When humans are confronted with new situations that challenge current understanding and practice they reflexively call upon their knowledge, culture, and history with their experience in relation to particular influences for constructing new ways of interacting with their changing environment. Knowing, being, and language constitute a foremost strategy in this process for understanding the problems being faced, possible options to enact, and communicating with others for support and advice. In general terms, pragmatic philosophy is the basis of integrated knowledge and inquiry learning around which many schools organise their curriculum and teaching. Paulo Freire (1972a) saw learning in a similar light, situated in community interest and being approached from a cultural standpoint. His literacy work in Brazil and elsewhere centred on 'culture circle' discussion among people who had come together to solve problems important to the community and who undertook a process of discussion, action, communication, writing, and reflection to produce a way of improving life conditions for the people.

Internationally, the specific recommendations of United Nations Declaration of the Rights of Indigenous Peoples (UN, 2007, p. 7) as a global consensus that took twenty years to assemble need to be considered, especially Article 14 which asserts:

1. Indigenous peoples have the right to establish and control their educational systems and institutions providing education in their own languages, in a manner appropriate to their cultural methods of teaching and learning.
2. Indigenous individuals, particularly children, have the right to all levels and forms of education of the State without discrimination. States shall, in conjunction with indigenous peoples, take effective measures, in order for indigenous individuals, particularly children, including those living outside their communities, to have access, when possible, to an education in their own culture and provided in their own language.

Decolonising knowledge for Indigenous peoples

All formal research is situated in the economic and political imperatives of the dominant society. This was particularly so for settler societies where European colonialism was the most extensive of all colonialisms and by the 1930s covered around 84 per cent of the world's land (Loomba, 1998, p. xiii). As more and more countries won their independence throughout the twentieth century, colonialism – defined as 'the conquest and subsequent control of another country and involves both the subjugation of that country's native peoples and the administration of its government, economy and produce' (Hiddleston, 2009, p. 2) – was markedly rolled back. Former colonies can however still be strongly influenced by the viewpoints, values, and practices of the colonisers that are very difficult to eradicate entirely from economic, political, bureaucratic, and cultural systems (Fanon, 1961/2005). For these reasons, the term *postcolonialism* has been used to describe the 'multifaceted effects and implications of colonial rule' through an understanding of and resistance to the continuing influence of colonialism from 'inauguration to the present day' (Hiddleston, 2009, p. 1). Colonialism has a similar meaning of conquest and invasion to imperialism, but the latter can incorporate different and broader strategies such as finance and market expansion, especially in the current period. For the Indigenous peoples of Australia and many other countries, the existence of 'postcolonialism' as defined above is very real as the struggle to reinstitute sovereignty and self-determination in all aspects of life continues. This is certainly the case where Indigenous peoples are involved in the dominant education system at all levels and its associated research procedures.

While dominant societies in colonial countries have generally disregarded and corroded Indigenous knowledge systems as a means of assimilation, there can still be opportunities for Indigenous and non-Indigenous peoples to work together on educational and research projects for mutual interest and betterment. In such societies, dominant research and knowledge paradigms remain in dispute with positivist and social constructivist methodologies vigorously contested. Drawing on market and accountability mechanisms, neoliberal ideologies emphasise quantitative rather than qualitative approaches and attempt to divide complex situations into small components for measurement. Under these circumstances, Indigenous and non-Indigenous researchers need to find creative approaches that respect Indigenous culture and ways of knowing and at the same time, are acceptable to the research profession. In the study on which this chapter is based (Dakich, Watt and Hooley, 2016) involving the use of iPads by mainly secondary school Indigenous children, a mixed methods methodology (Creswell, 2014) was used involving teacher comment and evaluations, student engagement survey, written literacy test for students, and state-wide student test results. Students were also involved in video production using tablet devices as a means of recording and generating cultural experience. Mixed methods methodologies however do not resolve the problem of how different data sets are analysed and interpreted, particularly when Indigenous perspectives are involved. In broad terms, we propose that the research methodology followed for the project consisted of a set of disconnected imperatives as shown in Table 9.1.

TABLE 9.1 Features of linear traditional research for combined Indigenous/non-Indigenous projects

Indigenous Community
• approach to knowledge • cultural, values context • trust and relationships • non-negotiable 'red line' issues • rights, responsibilities
Research Requirements
• funding obligations • time and timelines • recognised procedures • shared understandings
Constraints
• postcolonial settings • project design • external data • researcher positionings

When working with Indigenous communities, non-Indigenous researchers need to ensure that discussions occur with local communities over extended time frames such that all cultural and knowledge protocols are understood and respected (Singh and Major, 2017). This is a two-way process between all participants. Trust needs to be earned by non-Indigenous researchers in discussing the nature of the project, how it might proceed, and any non-negotiable red line issues that will prevent the project proceeding. The community for example might require ongoing daily participation in school curriculum that schools consider to be inappropriate. From a research management point of view, the research team will most likely have strict timelines to adhere to and may not be able to allocate the time to the project that they would prefer. Given the discussion above regarding postcolonialism, the project may be sited in organisations that have very narrow understandings of culture and knowledge, be under pressure to accept external data that is thought to be not appropriate to the research which in general, makes it very difficult for researchers who are aware of cultural and community protocols, but are constrained in their application.

We are aware that Indigenous researchers have identified a number of principles that they recommend should infuse Indigenous research. These include holistic epistemology, story, purpose, experiential understandings, tribal ethics and ways of gaining knowledge and an overall consideration of the current colonial relationship (Kovach, 2009). This becomes very difficult when working within formal academic arrangements that are not Indigenous (p. 43):

> How are we customising our Indigenous frameworks to fit within our tribal paradigms while communicating our process to Western Academia? And how is the language of frameworks itself ultimately chipping away at our philosophies? Can we carry out tribal-centred research within the academy without this framework language?

Based on the broad understandings of this protocol for researching of educational issues with Indigenous communities, we now propose the use of mixed methods within a democratic Community Narrative Research Model (CNRM), detailed as follows:

- Methodology: Community Narrative (as structured storytelling and knowledge) – community participation throughout to organise and shape knowledge production.
- Method: Mixed Methods – range of data sets to ensure that community can express knowledge production to fullest extent, depicted in Figure 9.1 (contrasted with Table 9.1).

Process: Consideration of central stories related to research question/s in continuing cycles over long periods of time denoted as identify, inquire, interpret,

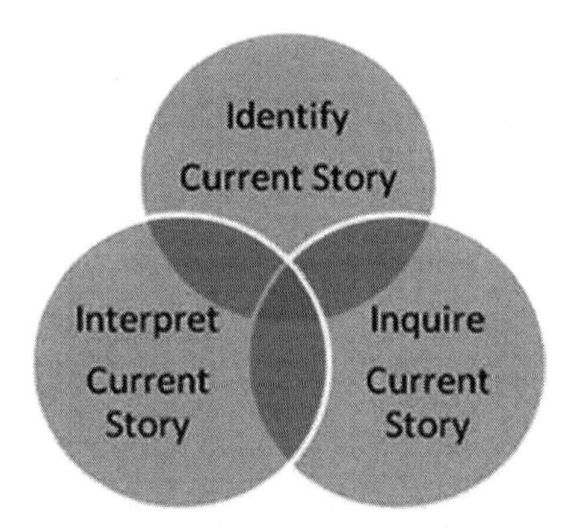

FIGURE 9.1 Features of cyclical, non-traditional research for combined Indigenous/non-Indigenous projects.

thereby creating new understandings for participants, or a new story and new cycle (see Figure 9.1). These cycles can be further clarified by research processes such as identify (describe current story, characters, ceremonies, issues); inquire (conversations around main issues and data gathering via mixed methods); and interpret (discussion of meaning, issues arising, creating new understandings of current story or recognition of new story for ongoing cycles).

To indicate how the proposed CNRM could operate, we will now link specific examples of data used in the project described in this chapter in an attempt to explain the need to ensure that data sources not only match the research questions, but also appropriately match the learning and cultural expectations of participants and the Indigenous community. As a supporting base to contextualise the identify level of the model, three sources of contrasting data types drawn from the project can be considered. First, requirements of the funding source for the project necessitated that National Assessment Program-Literacy and Numeracy (NAPLAN) (ACARA, 2017) scores for the Aboriginal and Torres Strait Islander students involved were included. The second data source selected by the researchers was derived from students' subjective self-reports of their perceived school engagement. Finally, evidence generated from interviews with community members was incorporated. These sources can be contemplated in relation to the inquire level of the model to substantiate their role as evidence for resolving the original research question. The NAPLAN data was not necessarily suitable in this context for supporting an inquiry into the development of Aboriginal and Torres Strait Islander students' school learning as an outcome of participation in an iPad program. The measure of school engagement, although not a direct reflection of school-based academic achievement, provided valuable insights into the need for students to feel engaged within their school community as a pre-requisite for personal and community learning. Community forums provided interview material that served as a more intimate overview of how those linked to the students within the programme perceived the possible learning consequences from the use of iPads. It was felt that this also provided the strongest cultural insight into the learning expectations and outcomes for the student sample.

At the interpretive level, the three types of data demonstrated varying capacities to contribute to the interpretation of the research question regarding possible connections between the use of iPads and learning improvement. Any change in NAPLAN score, positive or negative, was not necessarily attributable to the influence of the iPad programme and as such was never likely to contribute to any subsequent understanding. A small positive trend in student engagement scores collected over a nine-month interval could be used as supporting evidence to reinforce that the iPad programme may also facilitate improved school engagement, and hopefully, over time improvement in learning outcomes. This data source could be considered as efficacious in facilitating subsequent interpretation. The qualitative evidence from community members provided a direct interpretation of how those participating in the project connect with the research questions. Information of this type is probably the strongest contributor towards supporting a coherent interpretation of the research question.

Contribution to storytelling/formation

To indicate how the proposed CNRM could operate at the level of storytelling and communication, we now provide three separate examples of conversation that could occur within each of the identify, inquire and interpret cycles. These are indicative illustrations only of possible comments by non-Indigenous researchers in relation to a science project.

Midden story. Identify *(describe current story, characters, ceremonies, issues)*

Scattered along the west coast of Victoria are many Aboriginal middens of shell, bone and charcoal. Depending on the location, the middens often contain parts of periwinkle, crayfish, crab, abalone and other shellfish. Small groups of Gunditjmara and Giraiwurung peoples would camp along the coast near various food sources during each season of the year. Sometimes they would move inland to escape the cold winds that sweep in from the south during winter, perhaps staying near rivers and lakes for fish and eels. Over long periods of time, Aboriginal people were able to read the weather and the signs that change was near. The Elders told stories to explain what the young saw and experienced, stories that contained deep meaning accumulated over centuries that opened up ideas for discussion rather than provided answers. Sometimes the clans would relate closely to various animals, or land and sky formations that they camped near and shared their existence.

History story. **Inquire** *(conversations around main issues and data gathering via mixed methods)*

This series of photographs from the archives of the local historical society show groups of residents a century ago on social outings near the river. Horse-drawn carts are the means of transport, while the thick scrub and tall gum trees along the river banks are plainly seen. Newspaper articles from the times also comment on extensive clearing of the land for the grazing of sheep and dairy cattle, let alone the development of industry, roads, and housing for nearby settlements. I know that connections with the land have always been central to Indigenous life and seeing the vegetation cleared in this way must have been difficult to accept, including any impact on the supply of fish and eels across the southwest. We hear of local stories from the community regarding early contact and talks with farmers and woodcutters but little detail as yet. We understand to listen carefully to what is said and not ask too many questions, to let the words find their own level. The elvers (young eels) still migrate and climb the waterfall as they have forever and the story of their life cycle we believe is part and parcel of local lore that we have yet to hear. Respect for the river and creeks is common to all who live close to the land.

Family story. **Interpret** *(discussion of meaning, issues arising, creating new understandings of current story or recognition of new story for ongoing cycles)*

My mother spent most of her time on the beach and I well remember her saying how she would begin swimming in November. This would still be chilly along the coast, but more difficult now with February being the hottest month. The seasons seem to be changing with hotter and more humid weather occurring later than before. It is said that having an intimate relationship with the environment has fashioned Indigenous identity over the centuries and that care for the land creates Indigenous being and consciousness. Thinking of personal awareness and being in this way connects us with the Indigenous child in the classroom when asked by the teacher to read and write about topics outside of direct experience. It would be better for schools to relate to local culture and stories heard every day about family, community, and the land and what these mean for understanding our place in the world. Of course, the stories have to be accessible and be told by community members who have the knowledge and responsibility. Teachers also need to be able to incorporate these insights into their programs. Our lengthy discussions and data begin to open up this possibility, but we have only scratched the surface so far.

It can be seen that this illustrative family story told within the identity learning circle of the research methodology, outlines an aspect of the cultural history of a particular community, details what the dominant society calls science (in this case, ecology and/or biology) and indicates the place of storytelling in the process of knowledge production. Such stories may not emerge immediately when researchers and communities meet and may only be told when elements of trust and respect have been established. The identify process enables researchers and community to clarify the issues involved in the topic being researched and the knowledge and assumptions that may often be hidden. The process of storytelling by both community and researchers continues over the complete cycle/s of identify, inquire, interpret, as all participants share their experience and perspectives and bring their cultural backgrounds to bear on issues and problems that occur.

There are indications in the comments above that the research project concerns an investigation of literacy in schools. The initial conversation however raises issues regarding science and the environment and observations that range over what schools know as ecology and biology. This encourages the non-Indigenous researcher to think about how local experience, history, and culture can be incorporated into the formal school curriculum as the basis for reading and writing. Inviting a community member to discuss the food sources of Indigenous clans who lived near the coast and rivers and lakes will initiate literacy and language-based projects for students. Consideration of Freire's work on literacy and respect for community culture and understanding could be introduced at this point for the research team to begin to theorise how both literacy and research can be decolonised.

According to van Maanen (1988, p. 127), ethnographic 'critical tales' are 'strategically situated to shed light on larger social, political, symbolic or economic issues', not necessarily to provide answers to specific issues and problems. Positioned therefore within culture as 'systems of meaning', Williams' (1989) argument that culture involves current and emerging meaning and direction for communities seems appropriate. Williams writes,

> These are the ordinary processes of human societies and human minds and we see through them the nature of a culture: that it is always traditional and creative, that it is both the most ordinary common meanings and the finest individual meanings (p. 4).

In this passage, he sees culture as a way of life that is ordinary or encountered every day, fixed yet dynamic. A culture of this type is relational with other key ideas in society, such as learning, art, democracy, and transformation, and, moreover, is not restricted to persons of wealth and privilege. The connection between culture and meaning raises research questions about epistemology, ontology, axiology, and methodology, also identified relationally in the conversation of Ek and Latta (2013) regarding how meaning is actually constructed in the human domain.

Indigeneity, Indigenous identity

From these considerations, the construct of Indigeneity and Indigenous identity have emerged as crucial themes from the study reported. The concept Indigeneity is used here in its global context and is taken to denote a consciousness or worldview or set of perspectives that are distinctively Indigenous. It has been mentioned previously that Indigenous philosophy involves in part an interconnected view of the world, belonging to the land, kinship relationships, family and community storytelling, and oral conventions of knowledge and learning by Elders and other community members. The lack of an articulated and shared understanding of an Indigenous worldview is perhaps the major problem and barrier to insight if productive and shared work at the cultural interface is to proceed for research projects and in mainstream schools and other organisations. On the other hand, the concept of Indigenous identity arises from community culture and practice that may have overlap with non-Indigenous routines and customs but has major differences as well. Art, dance and song for example may be similar to both peoples but may be more connected to historical storytelling in Indigenous culture. A more intimate and spiritual connection with the land may be central to Indigenous meaning. Bridging the concepts and practice of Indigeneity and identity with the reality of non-Indigenous schools and the formal curriculum can be guided by '8 ways' (Yunkaporta and Kirby, 2011) whereby 'Teaching through Aboriginal processes and protocols, not just Aboriginal content, validates and teaches *through* Aboriginal culture and may enhance the learning for *all* students'. In addition, the ontological and epistemological ways of being, knowing, valuing and doing listed earlier, can be investigated across the common ground between mainstream and Aboriginal pedagogies, suggested as:

- learning through narrative;
- planning and visualising explicit processes;
- working non-verbally, with self-reflective, hands-on methods;
- learning through images, symbols and metaphors;
- learning through place-responsive, environmental practice;
- using indirect, innovative and interdisciplinary approaches;
- modelling and scaffolding by working from wholes to parts; and
- connecting learning to local values, needs and knowledge.

As the above points indicate, incorporating Indigenous identity into the school curriculum is not the same as seeking to Indigenise the school curriculum. Rather it involves a specific approach towards knowledge and learning that builds upon local culture and experience and that can be applied in all subject areas. As such, it constitutes an approach to literacy and to schooling engagement that is essentially epistemological in character, but which may generate some tensions with current pedagogies. For example, more experiential modelling and scaffolding by working from wholes to parts may not be regular practice in either literacy or numeracy, where more inductive and step-wise techniques may be preferred, including the use of ICT and various tablet stratagems. Learning needs to involve the whole child, family and cultural connections such that holistic meaning becomes available to enhance identity. Under these conditions, it may be possible to begin with small projects or pilot studies that provide experience for teachers and students in mind-sized bites and from which progress with learning can be evaluated. Such work may not result in an immediate epistemological paradigm shift, but it may mean that taken-for-granted non-Indigenous approaches will be challenged to some extent making greater participation and inclusion possible. Personal experience of the cultural interface and liminality discussed above then becomes accessible to frame further change and improvement.

At this stage, it is important to reflect again on the relationship between these key ideas of cultural interface and liminality and the notion of the public sphere raised by Habermas. In their discussion of language, learning, and the impact of digital technology, Gee and Hayes (2011, pp. 121–131) detail what they call 'three social formations'. They suggest that the 'oral social formation' allows for interpretation that is 'dialogic, interactive and flexible'. Next, the 'literate social formation' enables records of previous exchanges to be kept and to provide reference points for future proposals. Such records are often decontextualised (across time) and are considered differently than in the ebb and flow of conversation. Gee and Hayes then describe the 'digital social formation' that allows the oral and the literate to be combined and negotiated by users. It is these features that begin to break down the roles of authority and institution and which potentially at least, can return citizens who may have been excluded to more respected and participatory positions. A 'digital identity' emerges. However, as processes of globalisation and technologising have continued the strength of the public sphere and relationships within the public have become eroded, with greater emphasis on the individual and the local. It may be however that digital and social media will tend to recover notions of community and public as the channels of communication and expression are recouped by the citizenry. For Indigenous communities, this possibility is significant as less formal, more conversational and culturally inclusive literacy is accepted, contact with family and community often dispersed is maintained and connections with the dominant society can be explored in ways not feasible before. In this fluid context that is still being worked through, Gee and Hayes ask prophetically whether 'modern social media (is) giving rise to new global publics' (p. 131) or to new forms of separation and isolation. For Indigenous communities and families, this question applies to the public sphere of school and to the public practice of literacy.

Finally, we note other work being conducted in Australia that illustrates how many of the principles described above can be applied at the university level. Kutay *et al.* (2012, p. 47) outline their 'Indigenous On-Line Cultural Teaching and Sharing' project that is developing a 'web repository of narratives from Aboriginal community Elders, Aboriginal students and staff at the University of Sydney', so that such narratives can then be 'embedded in relevant scenarios within online, single-user interactive games to teach about kinship'. It is intended that the materials will support 'different professional learning contexts such as law, social policy, health and education'. Respecting community narratives and being encouraged to build scenarios that embody them is an approach towards learning that is congruent with the philosophy of Nakata and Yunkaporta and a process that can be supported by ICT across the curriculum. Enabling different worldviews to co-exist around the big ideas and contestations of the day is a major contribution to social progress that formal education pursues and one that must include Indigenous culture and knowledge. Looked at in this way, Indigenous identity becomes a crucial factor in comprehending Australia or nationhood itself and knowledge production. Although there may be differences in conceptualising time, space, and origins, these do not prevent counterviews entering perhaps tentatively into a harmonious relationship and establishing the basis of new knowledge, values, and satisfaction. Rather than being an added ingredient, Indigenous identity should be considered as a reconciling democratic construct of learning and 'systems of meaning' for all citizens regardless of social class, cultural background, or creed.

Acknowledgement. This chapter is based on the research and publication of Eva Dakich, Tony Watt and Neil Hooley. (2016). Reconciling mixed methods approaches with a community narrative model for educational research involving Aboriginal and Torres Strait Islander families, *Review of Education, Pedagogy, and Cultural Studies*, 38(4), 360–380, DOI: 10.1080/10714413.2016.1203683. I am most grateful that approval for publication of extracts from the paper in this book has been granted by permission of publisher Taylor & Francis Ltd, www.tandfonline.com.

Case 9. Apology for past wrongs

Jenny often wondered why she attended staff meetings. Sometimes the reports that were given provided a modicum of information, but generally not. Usually staff had to endure what had been decided elsewhere and how they were expected to comply. 'What did you think about that Indigenous education issue,' asked Melonie as they walked out of the meeting room, 'I'm not sure what we are supposed to do about that?' Jenny had always agreed that it was an important problem that teacher education had to deal with, but the discussion seemed to go around in circles.

> Well, I know the guidelines list content that teachers are expected to include in their subjects, but it just seems like more information to me, especially if we don't have an understanding of what it means to Indigenous communities. If it is only a matter of content, we would have solved the problem years ago.

As enthusiastic young primary school teachers, Jenny and Melonie were keen to ensure that their teaching was inclusive of all students and, in the midst of a multicultural suburb, connected with the background of all students and families. They had been disturbed however by a recent comment from their colleague Barry who stated that Australia and similar countries with Indigenous populations could not be an ethical, democratic and respectful country without resolving the question of the land. By this he meant recognition of Indigenous ownership of the land, rather than being taken by proclamation of the British when they arrived in the 1770s. From her reading, Jenny knew that colonisation had happened in many places around the world and that where land had not been returned, it remained an open wound. Barry had a long-interest in these issues and often mentioned that at least some progress had been made with what were called the 'Stolen Generations'. This was the practice in a number of colonised countries to remove Indigenous children from their families and more often than not, place in missionary or boarding schools a long way away. Many schools had watched on television in 2008 when then Australian Prime Minister Kevin Rudd had apologised for such an appalling policy. Mr Rudd said in part (National Apology to the Stolen Generations, by the Australian Parliament, 13 February 2008; extract):

> We apologise for the removal of Aboriginal and Torres Strait Islander children from their families, their communities and their country.
>
> For the pain, suffering and hurt of these Stolen Generations, their descendants and for their families left behind, we say sorry.
>
> To the mothers and the fathers, the brothers and the sisters, for the breaking up of families and communities, we say sorry.
>
> And for the indignity and degradation thus inflicted on a proud people and a proud culture, we say sorry.
>
> We the parliament of Australia respectfully request that this apology be received in the spirit in which it is offered as part of the healing of the nation.

Many teachers cried that day, the apology had been such a long time coming, but not a lot had happened since. 'It was a necessary step towards ultimate reconciliation between us,' said Barry, 'but there has been little improvement in health, housing, employment and education for our Indigenous brothers and sisters, including the kids we teach'. Jenny accepted that she could never truly understand how Indigenous people related to the land, whether the yellow sand of the immense coastal regions, the red sands of the dry outback deserts, or the expansive bush landscapes where kangaroos roam. She could appreciate that when that relationship was taken away, then people would lose heart and die. There had to be some way of coming together to understand each other and to make her classes meaningful and substantial.

10

FEMINISM AND EDUCATION

> There are no regrets only dreams unfulfilled
> dunes of disappointment challenging each step
> while the contradictions of precedent and present
> make for a constant reconstruction of meaning,
> along the water's edge and throughout one's mind
> a collage of imaginings shapes what is to become
> at sunset, a child wonders at impressions in the sand.

Similar to Indigenous philosophy and knowing, feminism offers a number of different perspectives to mainstream knowledge production. Feminism became one of the most influential movements of the twentieth century and, like other movements, continues to exhibit different strands of thought. In general terms, if I could suggest that feminism is taken to mean 'the advocacy of women's rights and the rights of all peoples on the grounds of political, social, economic and intellectual equality', then there are significant implications for education. For the purposes of this book, feminism should contribute understandings that are consistent with and support democratic intersubjective praxis strategies and which build upon the socio-cultural ideas and practices of all participants. If it is possible to identify a number of feminist knowledges within this context, then it could mean that women understand the world differently to men, substantially or in part and that their approach to knowledge production itself is also different, substantially or in part. Could this mean a feminist praxis or a praxis feminism? There is a literature that considers the relationship between pragmatism and feminism, with Hypatia of Alexandria (400BC) and Dewey's colleague and Nobel Prize winner Jane Addams usually mentioned here. These and other feminists will be discussed below. However the connection between praxis and feminism has not been explored extensively. Many women have made important contributions to philosophy and education, but their work is often overlooked or given cursory attention. This chapter attempts to identify feminist perspectives of praxis, feminist categories of knowledge and implications for education systems that will counter neoliberal imperatives of individualism, exclusion and superiority.

As a beginning point of discussion, we could propose tentative feminist knowledge and perspectival categories of the type:

- Activist: knowledge of social action.
- Archivist: knowledge of family and community.
- Educator: knowledge of the world and culture.
- Producer: knowledge of labour and production.
- Connoisseur: knowledge of discrimination.
- Womanly: knowledge of gender and difference.
- Sexual: knowledge of reproduction, children and practices.

Nominating a number of feminist knowledges in this way does not 'essentialise' women in terms of categories that necessarily determine what it means to be woman. Previously I have written that the social act 'to knowledge' could be understood as involving:

> human action of engaging, reflecting and interpreting the world. This is a dialectical social process of collaborative experience, where all of us of different cultures, ages and positions relate and interrelate with the world in a human way as we construct and reconstruct what it means to be human.

It is clear from this passage that I consider all humans as interacting with the world in a 'human way' that is based on our biological arrangement of nervous system, circulatory system, sensory systems and the like, mediated by our social experience. According to Plato in *Timaeus*, order is brought to the universe by a 'demiurge' or primary god, who acts as a craftsman in producing a good and beautiful cosmos. In this sense, Plato's essentialism provided for ideal forms to which humans could only approximately understand or experience, such as the ideal circle or triangle. Humans could have the idea of the ideal or essential form, but these needed to occur in relation to human practice. There is no intention here to 'essentialise' women from a dominant male point of view, to relate women to an 'ideal form' and, it follows to then relegate or consider deficient if seen to be unsatisfactory or flawed. In contrast, consideration of a 'human way' of being human has led to my notion of 'to knowledge' illustrated in Table 10.1.

From an epistemological point of view, Table 10.1 shows that all humans as knowledgeable actors approach knowledge in this way, but each will act according to their specific social and cultural experience, the material condition. In terms of the seven knowledge categories suggested above, they can of course be applied to how men interact with the world, but as a generalisation, it is suggested that the tendencies are more significant and dominant with women. From an ontological

TABLE 10.1 Indicators of social act 'to knowledge'

Social Acts	*Indicator 1 Curriculum*	*Indicator 2 Pedagogy*	*Indicator 3 Assessment*	*Indicator 4 Research*
To Knowledge	Participation with practice-theorising	Holistic, enabling culture, history, language experience	Narrative accounts of projects and processes undertaken	Investigate knowledge as process, not knowledge as given

point of view, the notion of 'being human' is not considered as a theological or religions question, where humans are created or designed by gods or demiurge, but as noted above, being human is a function of biology mediated by evolution and human experience. This argument was made clear by Marx and Engels (1888/1969, pp. 100–101) when they wrote about the abolition of the family happening when the exploitative role of capital was also abolished together with the corresponding abolition of the role of wife:

> The bourgeois sees in his wife a mere instrument of production. He hears that the instruments of production are to be exploited in common and, naturally, can come to no other conclusion than that the lot of being common to all will likewise fall to the women. He has not even a suspicion that the real point aimed at is to do away with the status of women as mere instruments of production.

This comment from Marx and Engels impacts on the question of pragmatism as it clearly locates women as participating fully in production, a process that has been markedly extended today. From this, we can ask whether there is a pragmatist approach and view towards feminism, or whether there is now a feminist view and approach towards pragmatism? In the former case, it is a straightforward matter to see a pragmatic approach being taken towards feminism as it is applied in various circumstances. For example, a part of the role of female trade union members will be to raise awareness of issues that impact on their employment specifically and to do this while being directly involved in all activities of their unions. In the latter case however the situation is not so straight forward, because having a feminist approach and view towards pragmatism indicates a feminist pragmatism, where pragmatism itself takes on a feminist character, where a different type of pragmatism evolves for application by all humans. I noted above that my proposed definition of feminism as 'the advocacy of women's rights and the rights of all peoples on the grounds of political, social, economic and intellectual equality', has important implications for education, the focus of discussion in this book. In relation to the possibility of a 'feminist pragmatism', it is now necessary to refine that definition in relation to education and to examine how that could be applied to the concept of 'to knowledge' detailed in Table 5.1. A working definition could be expressed in the following terms:

> Educational feminism involves social acts taken to advocate and support the democratic rights of all women and girls on the grounds of educational and epistemological equality.

This definition enables consideration to be given to how feminist character can or could be applied to pragmatism within formal and informal educational settings at school home, or in the community. Table 10.2 provides possible connections between the seven knowledge categories nominated above with the indicators of 'to knowledge', the process of knowledge construction via dialectical emergence. The table is for discussion only.

It is apparent that each of the comments in the cells above could apply to either male or female participants, but each is considered to be undertaken with examples from a feminist perspective. Significantly, the table demonstrates pragmatist feminism, rather than a feminist pragmatism. That is, the features of pragmatism remain the same, but are applied with feminist understanding and perspective that can be substantially different to a masculinist approach. It is entirely possible to imagine a classroom of any subject area that is pragmatist in approach but having each of the examples above being worked through from a feminist point of view. In general terms, this consolidates the view that all humans engage and interact with the social and physical worlds from a human standpoint, so that rather than thinking of pragmatism being feminist or masculinist in character, we can envisage a broadly humanist pragmatism for all citizens regardless of age, culture and socio-economic background.

In discussing the connections between pragmatism and feminism, Seigfried (1996, p.18) comments on the lack of literature in this field and the somewhat invisibility (to most) of female pragmatists. In her explanation, she notes:

> It is sometimes incorrectly assumed that pragmatism is missing from formal classification of feminism because it continues to make liberal assertions about the isolated individual, because it advocates the public-private split, or because it is scientistic. Richard Rorty's neopragmatism gives some substance to the first two assumptions, but he has also been criticised by other pragmatist philosophers for rejecting, among other things, the social and political dimensions of the pragmatist tradition. A more likely hypothesis is that the ascendency of logical positivism after World War II eclipsed pragmatism for reasons that feminists would reject. Pragmatism never disappeared, it was marginalised.

These reasons for the apparently weak connections between pragmatism and feminism ring true. I have mentioned previously that Dewey's work could be criticised – incorrectly in my view – for not giving more specific attention to

TABLE 10.2 Possible feminist social acts and perspectives with educational intent

Social Acts of Feminist	*Indicator 1* *Curriculum*	*Indicator 2* *Pedagogy*	*Indicator 3* *Assessment*	*Indicator 4* *Research*
Character	*Participation with practice-theorising*	*Holistic, enabling culture, history, language experience*	*Narrative accounts of projects and processes undertaken*	*Investigate knowledge as process, not knowledge as given*
Activist	Policy development	Integrated knowledge	Community Projects	Action Methods
Archivist	Community consultations	Access to documents	Interviews with participants	Thematic methods
Educator	Incorporates local and global	Identifies cultural events, artefacts	Comprehensive not superficial tasks	Combination of different data
Producer	Identifies place of women in work	Ensures knowledge linked to production	Ensures work experience is included	Analysis based on practice, theorising
Connoisseur	Framework of social justice	Equitable situation for all students	Range of approaches for different knowing	Multi-methods for data and analysis
Womanly	Includes female perspectives	Physical and metaphysical understandings	Descriptive, exploratory, open-ended	Gender factors for analysis
Sexual	Women's role, child rearing when appropriate	Showing personal relationships, connections	Incorporate tasks when appropriate	Emphasise knowledge from practice

collaborative factors of thinking and learning and that his notion of inquiry could be seen as individualistic. Mead, Vygotsky and Freire discussed the social and cultural basis of knowing in detail and their work complemented that of Dewey in a cohesive whole. Scholars of today and throughout the twentieth century should be aware of all features that contribute towards a theory of knowledge, but events such as fascism and war would make this exceedingly difficult. As well as logical positivism, the appearance of postmodernism and post-structuralism during the second-half of the century tended to swamp and distract philosophy. What also needs to be highlighted however is that feminism itself was also distracted by such events and that its position as a major social and political movement around the world was not regained and did not take place until the period of post-war reconstruction and the determination to create something better for everyone after the years of economic depression and destruction. What is generally referred to as 'First Wave Feminism' is taken to have occurred during the approximate period 1850–1930 and focused on the right of women to participate in all aspects of society, particularly the right to vote and to be members of the decision-making structures of society. The suffragette movement was central to this period. 'Second Wave Feminism' from 1960s–1980s extended participatory rights to those concerning the workforce, equal pay for equal work and sexuality and reproductive concerns. Defining the current era as 'Third Wave Feminism' is difficult given that there are different groupings with different emphasises including the family, economy, environment, civil rights, intersectional, online, sexual, postmodern and post-structural. This may reflect the dominant influence of international and national neoliberalism and the difficulty that all social movements have found in defining their location on the political spectrum (Wilkins, 2012). Our discussion above however indicates that 'Third Wave Feminism' needs to take education into account and to be able to articulate a philosophical position that is in the democratic interests of all citizens, male and female alike. In her book, Seigfried (1996, p.18) comments that 'it has sometimes been claimed that all feminists are pragmatists', a notion that fits well with our discussion so far. In order to make a better world, feminists, like us al need to think and act, to change what exists in systematic ways that have realistic prospects for success. To consider how some women have attempted this challenge in the past, we now turn to a small number of cases that provide precedent as well as inspiration.

Feminist scholars and activists

Any list of candidates for feminist pragmatism will be incomplete and controversial. It is also understood that for all progressive people around the world, the current era of neoliberalism makes social activity and the realisation of change difficult in every field (Connell, 2013). In Table 10.3 a small list of women has been nominated not necessarily for their background as committed feminist pragmatists, but for their dedicated work, action, achievements and mistakes in their respective fields that support women's rights and humanistic principles generally, work that is supportive of a pragmatist approach around the world. It is not intended that the

TABLE 10.3 Feminist philosophers, scholars and activists

Feminist	*Dates*	*Country*	*Activity*
Hypatia	350–415 CE	Roman Empire	Philosophy, astronomy, mathematics
Mary Wollstonecraft	1759–1797	England	Writing, philosophy, education
Emmeline Pankhurst	1858–1928	England	Politics, suffrage
Vida Goldstein	1869–1949	Australia	Politics, suffrage
Jane Addams	1869–1935	United States	Philosophy, reformer, suffrage, peace
Raden Adjeng Kartini		1879–1904	Indonesia Education
Kath Walker	1920–1993	Australia	Poet, reformer, Aboriginal rights
Maya Angelou	1928–2014	United States	Writer
Sally McManus	1971–	Australia	Trade unions
Malala Yousafzai	1998–	Pakistan	Education, activist

candidates should be categorised as 'Early, Current, or Post-Feminist', but that their work be considered on its merits. We can learn much from their experience and example. At this point, it should also be noted that regard for praxis feminism and feminist praxis will be taken up in Chapters 11 and 12 as knowledge production is examined in more detail. Feminist approaches to formal research and the construction or rearrangement of knowledge will be discussed below.

- Hypatia of Alexandria was a philosopher, mathematician and astronomer who became director of the Neoplatonist school of philosophy in Alexandria 400 CE. According to Watts (2017), philosophers had no formal authority in the Roman world of the time but had considerable influence. Watts suggests that Hypatia was 'a gifted philosopher who went into spaces usually dominated by men, taught ideas usually expressed by men and exercised authority usually reserved for men' (pp. 4–5). In 415 CE and because of her support for Orestes, the Roman prefect of Alexandria in his dispute with Cyril, the bishop of Alexandria, Hypatia was attacked by a mob of Christian monks and murdered.
- Mary Wollstonecraft was an English philosopher, writer and an early advocate of the rights of women. Together with her sister and other colleagues, she opened a school at Newington Green in 1784. She published *A Vindication of the Rights of Women* in 1792. She notes in the Author's Preface to *Maria or the Wrongs of Woman* that:

> The wrongs of woman, like the wrongs of the oppressed part of mankind, may be deemed necessary by their oppressors, but surely there are a few who will dare to advance before the improvement of the age and grant that my sketches are not the abortion of a distempered fancy, or the strong delineations of a wounded heart.
>
> (Wollstonecraft, 1994, p. 8)

Mary died at the age of 38 years, a few days after giving birth to her daughter, also named Mary (later Mary Shelly, author of Frankenstein).

- Emmeline Pankhurst was an English political activist who fought for the right of women to vote in parliamentary elections and was a leading member of the suffragette movement. She supported militant tactics to get the message across to the public and was imprisoned a number of times. As described by (Purvis, 2002, p. 7), 'Emmeline Pankhurst was a woman of extraordinary beauty, a fighter for the women's cause in which she so passionately believed, a charismatic leader and speaker who inspired the fiercest devotion and charmed most of those who heard her'. The British Parliament granted women limited suffrage in 1918 and full voting rights in 1928 shortly after Emmeline died. Emmeline Pankhurst has featured regularly in British newspaper polls regarding the most important women of the twentieth century.
- Vida Goldstein was a prominent Australian suffragette and supporter of human rights for all peoples. Australian women were granted the vote in federal elections in 1902 and after Federation, the vote in various state elections soon thereafter. In 1902, Vida travelled to the United States of America to speak at the International Women's Suffrage Conference, was elected secretary, gave evidence in favour of female suffrage to a committee of the United States Congress and attended the International Council of Women Conference. She was unsuccessful in standing for election to the federal parliament, but consistently supported the principles of compulsory arbitration and conciliation, equal rights, equal pay, the appointment of women to a variety of official posts, and the introduction of legislation which would redistribute the country's wealth. She was outspokenly opposed to capitalism, supporting production for use not profit, and public control of public utilities (see Brownfoot, 1983).
- Jane Addams was an American philosopher, social reformer, supporter of women's rights and peace activist. She established Hull House in Chicago as a refuge for women, children and refugees and worked closely with John Dewey in relation to education and the significance of social practice in learning. Dewey and Addams agreed that learning occurred through action, was communicated by 'being seen' and 'that people had to get to the point where they can feel together and act together' then 'think together' (Durst, 2010, p. 19). Jane strongly opposed the involvement of the United States in World War I and subsequently won the Nobel Prize for Peace in 1931.
- Raden Adjeng Kartini was an Indonesian activist and supporter of education and rights for women and girls (Connell, 2010). She is regarded as one of the

country's national heroes and a pioneer in the emancipation of Indonesian women. Kartini was the daughter of a Javanese nobleman who worked for the Dutch colonial administration and was exposed to western ideas when she attended a Dutch school. When she had to withdraw from school because she was of noble birth, she corresponded with Dutch friends telling of her concern both for the plight of Indonesians under colonial rule and for the restricted lives of Indonesian women. She married in 1903 and began a fight for the right of women to be educated and against the unwritten but all-pervading Javanese law. In 1903 she opened the first Indonesian primary school for native girls that did not discriminate on the basis of their social status. The school was set up inside her father's home, and taught girls a progressive, western-based curriculum. To Kartini, the ideal education for a young woman encouraged empowerment and enlightenment. She died in 1904 at the age of 25, after the birth of her first child. Her letters were published in 1911 under the title, *Through Darkness into Light* (see Kartini, 2017).

- Kath Walker was an Aboriginal Australian known by her tribal name as Oodgeroo Noonuccal and was a poet, activist, artist and educator. She was a major campaigner during the successful Australian referendum in 1967 to give citizenship rights to Aboriginal and Torres Strait Islander peoples. Oodgeroo served in the Australian Women's Army Service (1942–1944). She published her first book of poetry, *We Are Going*, in 1964, going on to become a trailblazer in published Aboriginal writing in Australia. Oodgeroo was Queensland State Secretary of the Federal Council for Advancement of Aboriginal and Torres Strait Islanders for 10 years in the 1960s and from 1972 was managing director of the Noonuccal-Nughie Education Cultural Centre on Stradbroke Island, Queensland. Throughout her life, she was a renowned and admired campaigner for Aboriginal rights, promoter of Aboriginal cultural survival, educator and environmentalist. She stood as the Australian Labor Party member for the electorate of Greenslopes in the 1969 State election. Although voting rights had only been in place 4 years, Oodgeroo decided it was time to '[s]how our black faces in parliament' (see The Australian Women's Register, 2018).
- Maya Angelou was an American poet, author and civil rights activist. She is well known for her 1969 memoir, I Know Why the Caged Bird Sings that became the first non-fiction best seller by an African-American woman. Maya recited her poem, On the Pulse of Morning at the inauguration of President Bill Clinton in 1993. During the 1950s, Maya had roles in Porgy and Bess, Calypso Heat Wave and released her first album, Miss Calypso. Maya was nominated for a Tony Award in Look Away and an Emmy Award for her role in the television series Roots. Maya was recognised for her style of autobiographical writing that encouraged descriptions of life without shame that supports African-American culture and opposes stereotypes. Her work is used in teacher education to illustrate how teachers can talk about race and rights (see Angelou, 2018).

- Sally McManus is the recently elected Secretary of the Australian Council of Trade Unions. She is the first woman to be elected to this position in the 90-year history of the council. Before becoming secretary, Sally was president of a state branch of the Australian Services Union and was leader of a major equal pay campaign. Her highest priorities now include growing union membership, winning stronger rights at work, taking on corporate greed and making sure that working people have stronger rights generally. She hopes that the balance can be tipped back in favour of ordinary Australians because so much wealth has gone to the top 1 per cent. During the neoliberal era, the decline of manufacturing in many countries and a corresponding shift to the services sector, has caused union membership to wane. Sally has therefore spent much of her short time as secretary visiting many work sites and union branches across Australia to rally support for fundamental changes to the laws that govern industrial relations and the employment conditions of workers. She is one of the new female faces of union leadership in Australia whose task is to change and reform national policy in the interests of the country as a whole and for millions of workers and their families (see ACTU, 2018).
- Malala Yousafzai is a Pakistani activist for female education and is the youngest Nobel Prize laureate awarded in 2014. She is known for human rights advocacy, especially the education of women and children in her native northwest Pakistan, where the local Taliban had at times banned girls from attending school. Her advocacy has grown into an international movement. At age 15, Yousafzai was injured in October 2012 by a Taliban gunman when he attempted to murder her. She remained unconscious and in a critical condition, but later improved enough for her to be sent to the Queen Elizabeth Hospital in Birmingham, UK. The murder attempt sparked a national and international outpouring of support for Yousafzai. In 2013, she gave a speech to the United Nations and published her first book, *I Am Malala*. She encourages education for girls and community action against illiteracy, poverty and terrorism. Yousafzai has commented that 'the terrorists thought that they would change our aims and stop our ambitions, but nothing changed in my life except this: weakness, fear and hopelessness died. Strength, power and courage were born' (see Malala, 2018).

What can be gleaned from these brief biographical notes regarding a very restricted list of women from around the world? Some tentative points can be made as a starting point. First, feminism is a continuing international and historical movement with extensive support that includes women of all ages, cultures and socio-economic backgrounds. There is a strong common core of priority including democratic rights within a particular country and rights for health, education and employment within a social context of improving literacy, family and personal resources. Second, while there is broad agreement on the features of First Wave and Second Wave feminism, the concept and practice of Third Wave and perhaps Fourth Wave are not generally agreed. What this means is that there are many women who are prominent in diverse areas of social and community life and who are united by their common dedication to rights, but who adopt different positions along a political spectrum. They have a broad and historical social conscience rather than narrow world view and relate to diverse issues. Third, there is a shared theme and commitment to action, that things must be done, that women have waited long enough, if improvements of substance are to be achieved. Much of this action may not be formally studied, theorised and documented whether verbally or in writing, but continues to mature and inform thinking among participants world-wide. Fourth and taking into account the different eras for each woman, it is difficult to ascertain whether a pragmatist approach, to some extent, is supported. Given the extensive activity of all those listed, there must have been widespread discussion among colleagues about the situations being faced, analysis of events and options and judgement about what should next be attempted, laying the basis for pragmatist understanding. Perhaps Dewey and Jane Addams were best placed in their considerations of Hull House, the Laboratory Schools at the University of Chicago and the most appropriate type of curriculum to implement for children. In broad terms, feminism can thus be conceptualised as a philosophy of practice located within naturalistic or social life situations that demand resolution, where not all participants need to be fully engaged in the same way all the time. It provides cautious evidence for Seigfried's statement above 'that all feminists are pragmatists'. It can be taken as a whole, a non-sectarian field of practice-theorising with major and minor discourses verbal and written and with contradictions of various types. Feminism continues to work itself out as occurs with all historical movements.

Feminism and research, the co-production of knowledge

Feminist criticism of formal research and modern science in particular has been a consistent thread of the feminist movement. Harding (1986, p. 9) for example in her early work commented 'The radical feminist position holds that the epistemologies, metaphysics, ethics and politics of the dominant forms of science are androcentric and mutually supportive'. In general terms, feminism supports a humanist approach to knowledge and research where, rather than the abstract knower whose background is irrelevant, the researcher is socially situated with social and cultural understandings that are brought to bear on the problems at hand. While rational and empirical observation, measurement and prior knowledge are important, they must be tempered by the voice of the researcher as research design, data collection, analysis and interpretation proceed. This raises the distinction between the context of research and the justification of results and whether male and female researchers by virtue of their sex go about these processes in fundamentally different ways. If we take Kuhn's notion of science taking place within the dominant paradigm of the era as a guide, then does it follow from a feminist perspective that men and women propose different epistemological paradigms to Marx's sociology, Darwin's evolution, Freud's psychoanalysis, Einstein's relativity, Bohr's quantum mechanics and Hawking's cosmology? Does a feminist explain nuclear reactions and the impact of dropping atomic bombs on

Japan differently to a male colleague? What is the feminist explanation for the function of the contraceptive pill? These questions are drawn from physical science and regard the justification of results rather than the context of research. In terms of the humanities and social sciences, we can refer to Virginia Woolf, Simone de Beauvoir, Benazir Bhutto, Toni Morrison, Germaine Greer and Angela Davis as examples of women who have acted and written passionately about war and aggression, discrimination and racism, poverty, politics, economics and social policy. Based on their experience and context as women, does it follow that the conclusions reached about violence, racism, war and aggression are different to those of men? Does it follow that a conservative woman will reach a conservative position on these issues, in a similar manner to a conservative male regardless of background? In other words, personal ideology that arises from intersectional experience and material conditions rather than sex and gender alone will dominate the standpoint, or justification of results, reached.

In summarising the determining features of feminism, Kohli and Burbules. (2013, p. 83) cite Weiner (1995, pp. 7–8) when she states:

> feminism has three main dimensions: political – a movement to improve the conditions and life-character for girls and women; critical – a sustained, intellectual critique of dominant (male) forms of knowing and doing; praxis-oriented – concerned with the development of more ethical forms of professional and personal practice.

These dimensions fit closely with the ideas that have been discussed in this book to date and which can form the basis of a feminist engagement with knowledge and research. It should be noted that these dimensions do not exclude male colleagues from adopting a feminist approach if that is congruent with their politics and ideology. Chapter 12 will discuss a humanist approach to life, education, knowledge and research, rather than either feminist or masculinist. In this regard, Weiner's first dimension seems uncontroversial from a humanist perspective as all aspects of social life need constant improvement for all citizens and specific action for specific groups is usually necessary. Dimension two needs to ensure that the critique of male forms of practice is critical rather than conservative and generates new challenges and practices without merely adjusting what is current. Defining an 'intellectual' critique is also important such that philosophical questions are pursued that extend into new paradigms of knowledge. Dimensions one and two are Second Wave considerations although, depending on how they are handled, can be included in Third Wave activity as well. It is dimension three with its emphasis on professional praxis and ethical conduct that opens new ground for Third Wave feminist influence on knowledge and research. Blackmore (2013, p. 3) for instance has pointed out:

> For feminists, research is praxis, in that theory and practice are interconnected and that any distinctions between theory/methodology/method are false. A feminist ethics and the purpose of equity underpins how one

> practices research. Feminist scholars have raised significant methodological and ethical issues around voice as well as ownership of knowledge and how women's symbolic and discursive representation impacts on women's sense of identity.

In recognising that praxis involves the close integration of theory and practice, Blackmore connects with the feminist holistic view of the world and of knowledge. She also notes the formal research distinctions between theory or knowledge paradigm, methodology and method that structure studies, but that this rigid distinction is false. Rather than accepting this predetermined rigidity, Blackmore contends that feminist research proceeds on the basis of an ethical and equitable understanding of the situation at hand. It should be remembered at this stage that the history of modern science involves not only a severance of the linkage between religion and philosophy but the initiation of new ways of going about knowing that did not rely on dogma. That is, the view that knowledge and truth were not only held by a small number of people in authority but could be approached by all citizens through their personal connections with the world, was revolutionary and changed the society forever. Within a general researcher understanding or paradigm of knowledge and appropriate methodologies for gathering data, a range of methods are possible, in fact a very wide range of methods that can often be used in different methodologies. It can be argued that science is ethically-neutral, concentrating on the problem, data and analysis in isolation, but the acceptance of a particular paradigm does not arise in a vacuum and must be based on social and cultural concerns whether recognised by researchers or not. While specific methods are adopted, the manner by which researchers think about the research is integrated, where paradigm, methodology and method are not separated. Dewey was clear in that method in its most general sense of how we learn, is not specified, but arises from experience as each participant engages with materials and problems, observes what is happening and judges next steps. He put it this way (Dewey, 1916, p. 179):

> Method is a statement of the way the subject matter of an experience develops most effectively and fruitfully. It is derived, accordingly, from the observation of the course of experiences where there is no conscious distinction of personal attitude and manner from material dealt with. The assumption that method is something separate is connected with the notion of the isolation of main and self from the world of things.

While I have noted above a number of commonalities that link feminist understandings and activities, there appears to be little and consistent in the feminist literature that takes a progressive philosophical position on epistemology and mind, a characteristic that I suggest should be incorporated into Third Wave standpoints. Fraser (Keddie, 2012) for example is one feminist writer who notes the distracting feature of identity politics. The cohesive and integrated view of experience and knowledge is not restricted to feminist thought and links nicely to the underpinning principles of pragmatism (Kaag, 2011). Dewey's note of the lack of 'conscious distinction of personal attitude and manner from material dealt with' is totally supportive of feminist perspectives today. A further gap in feminist theorising at present may be the omission of any continuing emphasis of dialectics, let alone materialist dialectics. However some examples of such connections are available (Hartsock, 1998; Russell, 2007). There seems to be a linear view of history with champions and villains, without cycles of experience and reflection that produce new thinking for ongoing application. Francis Bacon may be criticised as the originator of a universal dominating masculinist scientific method, a scientific method that seeks to control the humanities and social sciences, rather than appreciating that history moves on, building on what used to be for new circumstances. This somewhat static view of history indicates a lack of experience with the physical sciences and the transformation and negation of the negation of physical and life processes. Science does not impose but is constantly exploring and changing its understandings. Have feminists not looked to the sky and wondered at the shapes and patterns of clouds in constant movement, one transforming into the other, even present as mist and water molecules until various forms of coalescing occurs? Have their hearts not leaped and been touched by the movement of a needle on a scale when a magnet is brought near a coil of wire? Lenin (1976, p. 276) would ask why does the universe act this way and how does one entity change into another? He would further advise:

> Genuine dialectics does not justify the error of individuals, but studies the inevitable turns, proving their inevitability by a detailed study of the process of development in all its concreteness. It is a basic principle of dialectics that there is no such thing as abstract truth, truth is always concrete.

These questions and suggestions are offered from the standpoint of historical materialism, to strengthen feminism as all citizens around the world act together to envisage and create what is better for everyone.

Case 10. Voice and visibility

Wednesday mornings were always a highlight of the week. Of all her students, Alessa found Year 7 the most enthusiastic and willing to try whatever she had in mind. Becoming involved with formal science and its procedures and equipment was often something new for students when they first came to secondary school, even if they had stereotypical views of science being conducted by men in white coats. She smiled as she thought of a film she had seen on the weekend about a group of Afro-American women who had worked as mathematicians on the American space program. They had a tough time to be recognised but their talent won through in the end. 'Can we come in miss?' Alessa was shaken from her reverie as her class arrived at the door. At this time of the year, they knew what to do, organising themselves in groups of four and continuing with their units from where they had finished last time. As a young teacher, Alessa was very aware that preparation was everything for successful lessons, so she was pleased to be able to stand near the front bench and observe children busy in their endeavours and taking responsibility for their learning. Most were working on a chemistry unit regarding mixtures that involved filtration. Young fingers often found it difficult to fashion the appropriate structure of filter paper and to make sure that the muddy mixtures of sand, pebbles and other garden refuse and litter that Alessa provided did not overflow the glass funnel. 'Look at this miss, we are getting clear water in the beaker, who would have thought that?' 'It is amazing Cindy, do you know that's how we filter water at the reservoir so it's fit for drinking?' Each unit had some more advanced activities that students could try if they had time. Alessa helped Cindy and her group collect a number of test tubes that contained colourless liquids. They needed to carefully and slowly mix about 15 millilitres of one solution with another, making sure that they kept a record of which were being used, solution A with solution B for example. It seemed that nothing happened on the first attempt, but with their second try a bright yellow substance was produced, streaking through the tube. 'Where did that come from miss?', asked Helen excitedly, 'how did we get that yellow stuff?' Alessa agreed, she never grew tired of seeing this experiment work, or any experiment really, a wonderful display of the universe at its creative best. 'Well, we haven't studied this process yet Helen, perhaps next time, but it must have happened when the two solutions mixed, don't you think?' They were interrupted by Cindy who had just seen a rich white substance appear when she poured another two solutions together. 'Look miss, it's completely different it looks more . . . lumpy . . . and what a nice white colour!' 'Why are they different colours miss?' said Maria and then after a slight pause, 'they must be different somehow'. This is exactly what science should be about thought Alessa as she looked at the expectant faces around her. 'Do you remember last week when we had a brief talk about the colours of the rainbow and the spectrum of light?' There were nods around the group, before Maria commented that she thought we saw different colours because of the way light was and how our eyes saw light. 'That's a great idea Maria,' replied Alessa, 'good thinking.' 'So it might mean that the substances you have just made interact with light differently to the two solutions that you used. Now why don't you try and filter off each substance and see what they look like out of solution. By the way, we call these substances precipitates, so you could look that up on our science web site as well.' As Alessa walked around the science lab to check the progress of each group, she knew that her generalisation about Wednesday mornings had just accumulated more important evidence. What a great bunch of students. Some key ideas and practices of science had been encountered – mixture, filtration, solution, precipitate, change and even spectrum – and it was marvellous how imaginative thinking occurred when students were active and challenged, the power of experiment. She was sure there would be interesting discussions around a number of kitchen tables tonight.

11

BRICOLAGE

Practitioner knowledge and research

> Nonsensical to judge how perspectives
> arrange the personal, to see and touch the sand
> and sky and to know not what lies obscured
> how history and culture observe totality
> and build a mechanism for meaning that
> guides tentatively, change the eternal variable
> subject only to the wavelength and hope
> of interpretation, a constant fashioning
> of vagueness from within and without.

In ground-breaking work, Kincheloe identified the French concept of 'bricolage' as a philosophical approach to learning, knowledge and research and a means by which the various features of human life and scholarship impact on knowing, the features of history, economics and the like. As mentioned in my previous book (Hooley, 2018), drawing on the work of Levi-Strauss (1968) and Denzin and Lincoln (2000/2017), Kincheloe argued for the conception of teacher and researcher as 'bricoleur' such that diverse methodologies and understandings are brought together in the act of inquiry and research. He recognised that the complexity of the bricolage required knowing well a range of knowledge disciplines and pointed out that the concept of theory needed a new vision as well (Kincheloe, 2005, p. 324):

> Since theory is a cultural and linguistic artefact, its interpretation of the object of its observation is inseparable from the historical dynamics that have shaped it. The task of the bricoleur is to attack this complexity, uncovering the invisible artefacts of power and culture and documenting the nature of their influence not only on their own scholarship but also on scholarship in

> general. In this process bricoleurs act upon the concept that theory is not an explanation of the world – it is more an explanation of our relation to the world.

This is no easy task and will probably require teams of practitioners bringing to bear their combined disciplinary know-how, philosophies and experience on difficult problems and ideas. A critique of both quantitative and qualitative methods in the social sciences and humanities would reveal issues of bias, accuracy, consistency and authentication as well as the lack of relationship between the objective and subjective that raise many troubling issues regarding the stability and trustworthiness of educational research. In this way, Kincheloe's commitment to establishing a new comprehensive, cultural, respectful, inclusive approach of knowing, of rigorous interpretation of experience and of theorising through the bricolage opened up a new vision of learning and schooling for teachers and students alike. His view of research as embodying a total human experience was constantly emphasised:

> Any social, cultural, psychological, or pedagogical object of inquiry is inseparable from its context, the language used to describe it, its historical situatedness in a larger ongoing process and the socially and culturally constructed interpretations of its meaning(s) as an entity in the world.
>
> (Kincheloe, 2011, p. 180)

Advocating the practice of research bricolage in this way has the advantage that it is highly respectful of diverse community culture and values. However it has the disadvantage of not yet having gained acceptance as a recognised research methodology and the validity of its outcomes. There is extensive debate in qualitative research generally regarding new methodologies and like all innovation, new approaches go through a process of trial, documentation, discussion and refinement over a period of time before being accepted into the pantheon. Given that the proposal for discursive knowledge and research bricolage is being advanced for those communities that work within mainstream environments such as regular schooling, then ways must be found of data gathering and theorising of findings that relate directly to systemic operation and enhancement.

Instances of the bricolage have been encountered in our discussions of Indigenous and feminist approaches to understanding the world. In each case, certain groups of people have perspectives that differ from accepted viewpoints. For example and according to Darder, Mayo and Paraskeva (2016, p. 3), 'The key entry point to learning for Freire, then, is the learner's existential situation, or lived experience.' Indigenous peoples relate their knowledge to the continuing connections with the land and respect the authority of knowledge holders such as Elders and other senior community members. As outlined in Chapter 10, storytelling is a key aspect of learning with those who hold knowledge judging when it is appropriate to make that knowledge known. Feminist understandings are seen to be located in the material conditions of experience and also value the processes of discussion as a means of sharing and forming knowledge. In both cases, there is an orientation towards the bricolage with an emphasis on oral history and storytelling via informal talk either in small groups or larger community settings. These approaches are often considered as being at variance with modern science methodologies that rely on the predetermination of established methods and means of analysis. As mentioned previously, criticism of scientific knowledge paradigms and research methodologies are often taken from the humanities and social sciences, rather than physical sciences. Two examples may suffice to show the difficulties that bricolage must confront.

Many stories from around the world exist regarding the cosmos and night sky. These range from astrology, to describing star and planet observations and arrangements in human and animal terms, to modern concepts such as black holes and quasars. However it is difficult to see how a scientist can take into account Indigenous and feminist perspectives when attempting to theorise and explain dark matter and dark energy, the nature of a negative charge on an electron, or the issue of origin itself. A similar argument could be made regarding the behaviour of bacteria and viruses especially if this concerns illness and epidemics afflicting children in villages and cities alike. There may be community remedies that are applied, but these may be a matter of chance and unsuccessful when new diseases are involved. It must be reiterated that according to Kincheloe above, 'bricoleurs act upon the concept that theory is not an explanation of the world – it is more an explanation of our relation to the world' and that this does give rise to ideas that are generally thought to hold under different circumstances. Reasons for this may be unclear in the beginning but can be clarified through experiment and discourse perhaps over long periods of time. It is entirely possible that our personal experience with rubber bands as a child informed our thinking about the string theory of matter. It seems that the flexible, imaginative and creative bricoleur will not only utilise whatever research methods are deemed appropriate to pursue a specific problem but will also bring a range of thinking and understandings together to analyse and interpret observations that are made.

Advocating the practice of research bricolage has the advantage that it is highly respectful of community culture and values. However it has the disadvantage of not yet having gained acceptance as a recognised research methodology and the validity of its outcomes. There is extensive debate in qualitative research generally regarding new methodologies and, like all innovation, new approaches go through a process of trial, documentation, discussion and refinement over a period of time before being accepted into the pantheon. As mentioned above, proposals for discursive knowledge and research bricolage are being advanced for marginalised communities so that data gathering and theorising of findings that relate directly to the enhancement of such communities must be found. For example, consider a group of community members who are concerned about improving the literacy and numeracy achievement of their children together with their engagement with schooling. They develop a practitioner research protocol with a team of academic researchers and utilising their inventory of cultural wealth set about detailing the literacy and numeracy experience of their children at home. This involves discussion with parents and Elders, talking with friends and family, attending local events and ceremonies, interacting at sport and hobby clubs, visiting work sites, browsing newspapers and magazines, utilising computer applications, journeys through cities and countryside, writing letters, notes and e-communications and visiting libraries, museums and galleries. From this catalogue, key features of social life and knowledge can be connected with a broad understanding of literacy and numeracy and how they connect – or do not connect – to main factors of school life and knowledge. Community experience can be documented in various written, visual, or oral form in a 'Funds of Knowledge' (Rios-Aguilar *et al.*, 2011) approach. Academics or others working as critical friends analyse community-school experience and can suggest explanations and other data to collect in a next cycle. A number of these steps are commonplace in recognised research, but the range of data collected and the relationship between participants are much more personal and non-traditional.

Capstone research investigation: exploring the bricolage

A capstone project or experience for university or college students enables participants to draw on the range of literature, knowledge and practices developed throughout their studies in an integrated fashion. They also open up the question of research itself. Such projects are well known in the United States and are becoming more familiar elsewhere. The name derives from the central stone of an arch or bridge from which other stones or bricks are located. At the undergraduate or postgraduate levels, a research or researchful project is generally taken as a culminating study towards the end of a formal program. Capstone projects can of course be conducted within the regular paradigms and methodologies of scientific research adopting traditional approaches to analysis and interpretation. However for the purposes of our discussion of research, capstone projects can be designed so that they differ from a minor or traditional thesis in that they encourage a creative, non-traditional, practitioner approach to knowledge production, with an emphasis on 'researcher as bricoleur', on personal reflection and the innovative presentation of data and findings (Ings, 2015). This is the notion of capstone that connects closely with pragmatism and constructivism whereby all understandings and key principles are brought together to support and investigate a particular issue of interest to the researcher. Students scaffold and conduct a research project that involves them in the definition of an appropriate problem, review of relevant literature and the design of suitable methodology. Through this process, students collect relevant data and suggest outcomes based on data interpretation and analysis. Working in small groups or learning circles, participants will construct a narrative of their research journey so that they can reflect on how the project has influenced their views of research, where knowledge comes from and how they can think creatively and critically about knowledge itself (Hodkinson and Macleod, 2010).

Students explore a nominated field of education that they regard as directly related to their personal learning and curiosity and by so doing, come to a deeper understanding of research methodologies and methods so that they might design projects based on their experience rather than at the direction of text books or teachers. The concept of 'becoming researchful and insightful' means adopting an 'inquiry habit of mind' that can be applied at all times whether formally or informally as a 'practitioner researcher' and as a 'research informed professional'. This approach involves therefore not only undertaking a research project, but by so doing, investigating the complex nature of research and how we pursue knowledge of complicated issues in novel and integrated ways. Creative data gathering, report formats and the expression of ideas are encouraged including music, video, art, poetry, multimedia and social media. There is a strong focus on democratic peer learning through the learning circle structure and the manner by which professionals contribute to mutual learning within their fields of inquiry. Based on these ideas, learning outcomes for capstone project researchers also need to encourage 'inquiry habit of mind' regarding research and knowledge and not be confined only to traditional project aims and research questions. Learning outcomes could be of the type:

- demonstrate researchfulness and insightfulness through the effective communication of ideas and concepts developed from the critical evaluation of research data;
- participate in discourses regarding the nature of research and researchfulness including critical and creative methodologies and methods;
- articulate understanding of how to ethically conduct research or a workplace investigation;
- critically review relevant and current scholarly literature/s relating to the investigation;
- analyse and synthesise a range of conceptual and empirical materials to draw defensible conclusions.

In contradistinction to the approach taken above, Hauhart and Grahe (2015, p. 5) report that studies of capstone projects in the United States indicate:

> First, the data show that capstones are widely offered across many disciplines and that, in most cases, those capstones are offered to students in their major fields of study and not as interdisciplinary or general education capstones. Second, these studies suggest that disciplinary capstones are most commonly research project capstones.

From other educational jurisdictions, the capstone projects suggested here are analogous to the minor thesis conducted within particular fields of study and usually with a somewhat restricted focus and time framework. For institutions that already offer a minor thesis pathway, it seems a little purposeless to either provide a similar approach under another name, or to include a study within a programme that is not dissimilar to other projects. Both types of projects, minor thesis and capstone, need to be substantially different in conduct and intent to ensure that staff and students can proceed with clear distinctions in mind. These distinctions will allow for capstone work, a wide range of data as mentioned above, the compilation of diary entries or similar notations to narrate the research process, collection of all materials in portfolio formats and the creative expression, analysis and interpretation of data that situates learning within social-cultural contexts. For example, outcomes can be expressed in story, poetry, video, posters, role plays, exhibition and social media, whatever is judged and justified by researchers as most appropriate to communicate the consolidated and tentative ideas encountered. For example, I have been present when capstone participants were brought to tears by a poem written by an undergraduate researcher describing how she found online learning programs oppressive. She felt trapped, within a prison cell, impossible to escape from the rigidities of institutionalised instruction. Without detracting necessarily from the recognised scientific method of research, although certain criticisms of narrowness are appropriate, the critical and creative approach to capstone projects described here does tender other avenues to knowledge that are more comprehensive and multidisciplinary than the strictly scientific.

Case writing and learning circles

Theorising that knowledge and research can be understood as 'intersubjective practice', or as engaging and understanding our changing relationship with the world, it follows that learning environments should enable all participants to consider issues of their interest, curiosity and imagination. This of necessity will draw upon background personal and community culture and will involve fertile language situations of conversation, description, initiative and creativity. There is thus a combination of experience, experimentation and interest as current ideas and practices are transformed into the new for all participants whether child, adult, or formal researcher. Accordingly, assessment and evaluation of research outcomes can be taken to mean practice analysis of learning environments and the identification and monitoring of new practices so created. In considering education and research in this way, we have drawn heavily on the work of Freire (1970/2002) whom, as we know, advocated a process of learning circles and community practices to support learning and to change social and educational conditions. This involved cycles of generative themes, problematisation or problem posing, democratic dialogue, problem solving and personal and community action on issues of concern and importance to participants. Under these circumstances it is possible to involve all those interested in meeting events (objects or contacts of life experience) as they occur and to agree on their resolution (how to proceed). Participant experience and culture needs to be respected as the starting point and the basis of all learning as current knowledge comes into conflict with new possibilities. Abstraction and theorising becomes possible as experience, reflection, discourse accumulates. This approach fits with the concept of intersubjective practice and the monitoring of practices noted previously.

Following Freire, the practice of learning circles can be an organisational principle for researchful capstone projects to enable the collaborative sympathies of group members to be realised. Case writing can also be a powerful means of documenting and narrating the project from which key themes can be extracted (Hooley, 2015, pp. 168–173). Defined as a data collection mechanism, cases can involve brief descriptive accounts of significant events during the research process. They should not attempt to answer any of the incidents involved but describe in a clear and unambiguous manner what has occurred. Researchers can then share their experiences in a 'case conference' format involving each member of a learning circle reading their case, followed by discussion and questions from other participants. Once all cases have been read and discussed, themes, tensions, issues arising and suggestions for ongoing work are compiled as researcher knowledge to that stage of the project. It is possible to include comments from a 'community or critical friend' at any time during the project, to provide more extensive experience, advice regarding problems and issues that arise and to assist with theorising. Depending on whether the project is undergraduate, graduate, capstone, or regular research being undertaken by students, teachers or other researchers, the case writing and conferencing process enables project and research monitoring to occur in relation to the type of thinking and knowledge that is being generated and located in the minds of the research team. If agreed as the purpose of the project, it can be written up as an exegesis, involving project artefacts (posters, videos, poems and the like), plus a written explanation of the artefacts and research outcomes.

Discussion in this section has attempted to indicate that the process of knowledge production is broad and can be regular and traditional in relation to what is generally understood by the term scientific (Gore, 2017). It can also be flexible, critical and creative in terms of project design, data collection and analysis and project reporting. Being 'researchful' involves an expansive set of learning outcomes that necessitate ongoing reflection (as per diary entries, notations and case writing) and a reflection and reflexive phase at the conclusion of the project. Appropriate starter questions at this stage could include (Capstone Writer, 2009, p. 71):

- Do you believe that your project was a success?
- What would you have done differently?
- What have you learned about knowledge and research from the capstone experience?
- Do you think the difference between research and researchfulness is useful?
- Can the notions of 'creative' and 'scientific' research go together?
- How would you define 'ethical' research?
- Could other projects or follow-up research benefit from your work?
- Are you a changed person because of your research project?

These are difficult questions and suggest a professional rigour to research whichever approach is adopted. First, while the bricoleur uses the most appropriate method at hand, the bricoleur must be aware of the range of methods available and which one is most suitable for the problem that is presented. What is the best angle for the spouting along the eves of a house to ensure that water will not overflow into the roof, or allow all water to run smoothly into the down pipe? What is the best approach to record viewpoints from members of the football club committee: interview, discussion group, statistical survey, video, or case writing? Second, determining the best approach is a judgement based on reading of the appropriate literature and the accounts given by others of similar situations. Will these plants exist in mildly arid conditions according to the latest nursery information? What do we know about urban children of Year 5 and their use of social media? Third, taking an independent stance towards what others say and being able to compare and contrast different viewpoints. Placing solar panels on the roof is a good idea, although there are few hours of sunlight at this location. These survey results from the local council seem reasonable, but it looks like only a small sample was taken. Finally and perhaps the most difficult of all, being aware of our own thinking and what experience we take into account when considering problems and making decisions. In each of the examples just given regarding what might be called informal and formal approaches to knowledge, we all make decisions about how we understand that knowledge based on our backgrounds, the intersubjective nature of knowledge as mentioned above. Exactly why we discern specific themes in data that are different to the discernments of others, is unclear. Exactly why some parents and teachers appreciate the unbounded energy of children, while others view this as a problem to be neutralised, cannot be totally explained. There will be indicators of course. In the same way, there will be differences in how we consider music, painting, literature, tennis and birds on the wing, even between those of very similar background, the haut monde or hoi polloi. Bourdieu (1984) described this in terms of 'taste' and 'distinction'. At some point, researchers will be in the same position, of attempting to make sense of all the data at their disposal in relation to their personal and community, cultural and economic frames of reference. This is so for informal and formal, scientific and creative research alike.

Research, bricolage and pragmatism

Pragmatism and bricolage as approaches to research go hand-in-hand. As mentioned at the beginning of this chapter, Kincheloe proposed an expanded view of bricolage from an item constructed from a diverse range of procedures and materials, to 'a philosophical approach to learning, knowledge and research and a means by which the various features of human life and scholarship impact on knowing, the features of history, economics and the like'. On the other hand, we recall from Chapter 1, that Peirce described pragmatism in the following way: 'Consider what effects, that might conceivably have practical bearings, we conceive the object of our conception to have. Then, our conceptions of these effects is the whole of our conception of the object.' Further, discussion in this chapter has indicated that human approaches to knowledge production involve a combination of informal knowing arising from personal and community experience, to formal knowing that adopts recognised scientific processes. The relationship between these understandings is depicted in Figure 11.1.

It is proposed that all four features shown in Figure 11.1 cannot be separated but are integrated and exist simultaneously. Taken together, the features do not suggest a research process to be implemented, but constitute an epistemological concept of action, thinking and knowledge production. Similar to the theory of 'orbital' envisaged as surrounding the nucleus of an atom, the four features can be imagined as 'shells' of energy that interrelate as humans live with and in their worlds. We can refer to the quantum notion of 'entanglement' here where the states of different particles intersect and cannot be described independently of each other. In educational terms, this means that we envision all participants as existing and acting at different positions within the shell, but as being connected to all other participants at the same time, dialectically. Mead's concept of being able to exist in different systems at once also comes into play here. For research purposes, there are two important implications. First, an individual or community working on a particular problem may be drawing primarily on personal practice, although connections will be made with all other experience. In this way, the experience and culture of social groups such as Indigenous, feminist, working class, ethnic,

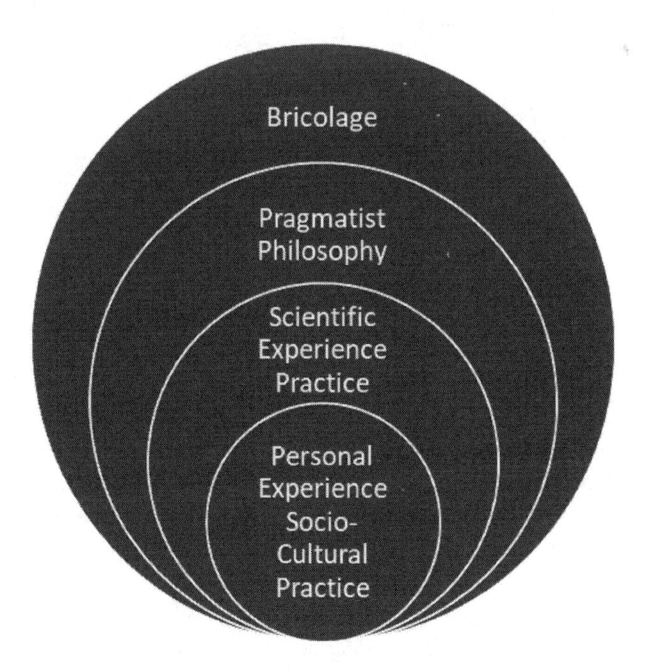

FIGURE 11.1 Relationship between bricolage and pragmatism

disabled, city and country is accepted as being central to new knowledge and that new knowledge is always being co-produced in relation to broader concerns of the bricolage. Second, the shell model provides an overview of all formal research projects, not as a method to be followed, but a conceptual map of where knowledge comes from and what key features must be taken into account before, during and after research projects proceed. Again, regardless of the research being undertaken, participants will need to consider the presence of different perspectives, Indigenous, feminist and the like and their impact on knowledge. For teachers and educators at all levels, the shell model provides a clear understanding of the complexity of classrooms and what is expected of staff and students as they grapple with ideas that confuse, challenge and interest and attempt to make sense of what society considers important and their own relationship with social experience. In the complexity and entanglement of classrooms, staff and students will be at different epistemological positions at all times.

It may be useful at this point to distinguish between the 'Action Research Spiral' and what has now been identified as the 'Pragmatist Research Shell'. In the esteemed trajectory of Lewin (Lewin and Lewin, 1948) and Stenhouse (1983), Kemmis and McTaggart (1988) proposed a spiral of project and research activity involving specific moments and overall cycles of 'plan, act, observe, reflect'. This work continues to be extensively cited in the educational literature as a significant movement towards educational practitioners being enabled to investigate their own practice and to make changes to benefit the learning of all participants (Kemmis, 2010). In association with the Action Research Spiral, the Pragmatist Research Shell provides a philosophical whole or entirety that completes the conceptualisation of practice-theorising for all human projects, formal and informal. It ensures that both elements of practice-theorising occur so that new know-ledge or fresh rearrangements of knowledge are produced and that one-sidedness that diminishes each, is overcome. Other correlations could be made here, such as Bourdieu's habitus-field construct, or Habermas and his lifeworld-systems hypothesis. There is a similarity between habitus-field and bricolage, although the inclusion of pragmatism as the guiding philosophy of the shell model affords greater connectivity. Lifeworld could also be considered as similar to bricolage, although making systems a separate imperative of rules and resources weakens if not misunderstands the complete lifeworld character as lived. All of these ideas however attempt to aid explication of human understanding and respect for knowledge that is located in the common weal of social experience.

In proposing the conceptual 'pragmatist shell' model of research and knowledge formation, there is a further question to review. It remains a difficult situation to describe, but in the process of critical pragmatist inquiry, humans will be thinking and acting at the same time in the process of dialectical emergence noted previously. This means that thinking and acting, theorising and practising are not separated, but exist in relationship between humans and between humans and objects of the world. Biesta and Burbules (2003, p. 59) consider this question when they cite Dewey (1938, p. 108) as commenting:

> Inquiry is the controlled or directed transformation of an indeterminate situation into one that is so determinate in its constituent distinctions and relations as to convert the elements of the original situation into a unified whole.

They go on to state that 'the process of inquiry thus consists of the co-operation of two types of operations: *existential operations* (the actual transformation of the situation) and *conceptual operations* (reflection or thinking)'. It is not possible to describe the process that a human goes through when a new idea forms in consciousness, or exactly why this occurs, except we know that it happens. We may 'know' that the amount of sugar in the chocolate cake can be varied to alter the degree of sweetness, but what we think about or how we evaluate 'sweetness' cannot be explained, how we make judgements about an indeterminate situation based on this property of 'sweetness'. Similarly, the classroom teacher can attempt to change the indeterminate situation found with a Year 8 class each day, but the judgements made either in preparation or in action, will often occur apparently spontaneously. Exactly the same takes place when considering data sets in research and the occurrence of a theme that suddenly comes to mind. What can be suggested from the 'pragmatist shell' concept, is that experience from all shells needs to be available and that the researcher should attempt to reflect on the relationships between all.

Questions of how humans come to know reality are questions of practice. While it is possible to study an object or idea, it is only through social practice that the object or idea becomes known subjectively. In structural terms, learning concretely refers to the assimilation of new conceptual models into current intellectual frameworks through a process of social practice undertaken by the actor or learner. Young children in particular are wonderful concrete learners, but these opportunities may be restricted by formal schooling as they become older. While it is possible for schools and universities to adopt a supportive view towards combining theory and practice, this is more often a contemplation of practice from afar that has been pursued elsewhere at another time and place. Different approaches to practice can of course be followed in different personal social and professional circumstances. For example, Kemmis (2012), in his discussion of practice, notes the different relationships between the individual, the social and a range of different objective and subjective perspectives. He identifies a 'reflexive-dialectical' relationship when' practice is socially and historically constituted and as reconstituted by human agency and social action, critical theory and social science'. Kemmis cites the work of Lave and Wenger (1991) and their notion of 'learning architectures'. He extends this concept to that of 'practice architectures' to include how both learning and the patterns of work are structured and the ways 'that practice is constructed, enabled and constrained'. In an Australian study aimed at investigating the key benefits and challenges of site-based preservice teacher education, five concepts of practice were identified (Eckersley *et al.*, 2011, pp. 86–89):

- inspiring practice;
- transforming educational discourse;
- challenging culture and pedagogy;
- investigating recursion;
- enabling ethical knowledge.

Each of these practices has a dual character, for example practice that is inspiring regarding the project at hand and practice that inspires further practice. It is noteworthy that the term 'recursion' is used in the context of educational practice. Recursion is an expression or procedure found in a variety of knowledge domains including mathematics, language and computer science, but for the study being undertaken, educational recursion was defined as (*ibid.*, p. 88):

> A self-referential system of relationships between key features of pre-service teacher education that constitute site-based learning and which involves each feature referring to all other features as experience is constructed and reconstructed and which establishes new understandings of each aspect of experience.

This definition could easily be applied to the Pragmatist Research Shell, where new aspects within the shell are created and establish new relationships with other aspects as the project continues (Miles *et al.*, 2016). As the social bricolage changes and impacts on socio-cultural experience, a changed relation with scientific experience will surface. Freire (1992, p. 88) makes a similar distinction between the knowledge of reality and the transformation of reality when he writes that the 'authenticity' of consciousness-raising 'is at hand only when the practice of the revelation of reality constitutes a dynamic and dialectical unity with the practice of transformation of reality'. This notion of an authentic consciousness could therefore be suggested as existing within the relationships of the shell model. Bourdieu (1977) considers thinking that is isolated from practice as a 'scholastic' matter used in an academic sense, particularly in defence of orthodox viewpoints. He discusses the philosophical notion of 'doxa' as common opinion existing between the subjective and objective and from which the orthodox and heterodox opinions arise. Bourdieu suggests that an adherence to tradition occurs when there is a 'misrecognition' of possibility and the natural order is taken for granted. This can occur in classrooms when teachers can mistake and misrepresent the capabilities of children as deficient and unchanging, an ill-informed common sense rather than an informed professional understanding. A comprehensive approach to research, teaching and learning as framed by the Pragmatist Research Shell that is critical, reflexive and dialectical provides analytical and conceptual strategies for constructing learning environments that assist all learners to investigate complicated and productive ideas and to make them their own.

Informal and formal education that sets about interacting with and transforming the objects of human experience, conceived as a matrix of interacting and referential personal and social practices, enhances dispositions of practice, knowledge and equity for all participants. In this regard, Cuban (2016, p. 138) notes the importance of researchers 'recognising and then integrating' what Peterson (1998) called their 'multiple identities'. Of course, connections within the shell of experience will not occur miraculously but must be secured between knowledge immersed in practice and knowledge abstracted from social practice. This is the task of educators, the practitioners concerned and the programme design authorised and is the key aspect of critique regarding conservative education systems. It is difficult to see how immersion in practice can occur when isolated from the confusions of social reality, although there are important times for discussion, reflection and analysis. It is difficult to see how participants can experience the wonderment, betterment and challenge of knowledge in its glorious extent by induction into skills alone, or indeed, by reflection alone. It is difficult to see how learners and researchers can investigate meaning without respect for their current cultural and social backgrounds, or make intellectual leaps across and within the bricolage and paradigms of knowledge, if personal understandings, emotions and aspirations are denied. Research and theorising that attempts to grasp our relationship with the world and its events is common to all peoples on earth steeped in distinctive knowledge, culture, language and practice.

Case 11. Country knowledge

Like most young teachers, Jim was up for anything. He had come into teaching with an activist background and had a genuine commitment to making a difference in the lives of ordinary families and children. As a mathematics and science teacher, he felt that he had a reasonable understanding of environmental issues from a scientific point of view and that he had a key role with the local landcare group and editing its monthly online magazine. Jim had also become involved with the education committee of his union that was working on a new curriculum policy with an integrated humanities and sciences emphasis. With this type of profile, it was logical that Jim would come under notice and he had been encouraged to stand for the union's state council. At one of his first council meetings, a senior colleague had nominated Jim to visit a country branch of the union to discuss a problem that had arisen in the town. After some months of negotiation, it had been decided that a wind farm with a number of large turbines would be built, some being close to the primary school. As he pulled up near the meeting hall, Jim took a deep breath, realising that a difficult couple of hours lay ahead. He was a strong supporter of renewable energy resources including wind as a measure against global warming, but he was there to discuss and represent the interests of the teachers who were, as he had been told, against the towers. Following some brief comment from a representative of the company involved, several parents and teachers spoke generally in opposition to the wind farm, although some wanted more detailed information. 'What about the red-tailed parrot?' a young mother and local farmer suddenly exclaimed, jumping to her feet, 'there are only a handful left now'. Jim was aware that there was a small flock of rare parrots in the area and two or three had been killed when they had flown into the turbine blades. This was a contradiction in his mind that he had put aside, but now it was staring him in the face. 'Perhaps it's time we heard from the union,' said the meeting chair looking pointedly at Jim, 'do you have a view on the red-tailed parrot?' As an inexperienced teacher let alone union representative, Jim had never been in this situation before and desperately tried to think through his confusion as he stood and gripped the chair in from of him. 'Well, we seem to have competing interests don't we,' he began hesitantly, 'the interests of the environment and the interests of the children at school and possible health effects.' 'To be honest, I probably haven't given a lot of thought to the situation of the parrot, so I'll be pleased to follow that up with you.' Jim paused, then seemingly out of nowhere it came to him. 'I have been thinking though that it might be possible to develop a solar energy plant rather than a wind farm, that would protect the school and the parrots.' There was silence in the hall for a few seconds before the meeting chair responded, looking at their representative in the front row: 'Yes, that was an early suggestion, but the company said it was not a viable option for them.' 'Well, be that as it may, but the technology is changing very rapidly and from what I have been reading in the latest science journals, the production of solar energy is becoming more efficient and cheaper each year.' Jim was encouraged by the mutterings and nodding of some heads in the room, so he quickly added, 'I'd be very pleased to sit on any working party that you established – together with the company of course – to investigate the feasibility and economics of changing from wind to solar, or indeed, a combination of each.' As he stood sipping his cup of tea and observing small groups of parents, teachers and some children discussing their views on how to proceed, Jim could not help feeling a little pleased. Not only was community concern for protection of the red-tailed parrot the tipping point of the meeting, but also his suggestion for changing the nature of the renewable energy involved had been accepted as a sensible and realistic compromise. He wondered what had prompted that idea in his head and made a mental note to himself to make sure that he kept up his journal reading each week. He would have to find time to sit on the working party, but it was a great opportunity to learn more about the environment – from the ground up.

12

DIALECTICS OF DEMOCRACY, CITIZENRY AND KNOWING

A letter to those who would act and teach

> Stretching uninterrupted to the horizon
> imagination spreads to fill the expanse
> of meaning overflowing the boundaries
> usually constrained by hegemony and cant
> yet when wandering off the beaten track
> encouraged by the liberation of immensity
> landscape shapes and intensifies thought
> embracing the contradictions of experience,
> what is worthwhile cannot be suppressed.

Dear Colleagues

Over previous chapters, I have attempted to write about philosophy and education as a philosophy of practice. They have been written with educational practitioners in mind, in effect meaning all humans as we engage each other and the world every day. It has not been possible to theorise philosophy itself and to compare different approaches such as between western (e.g. Greek) and eastern (e.g. Chinese) understandings (Peters, 2015). Hopefully a narrative style with examples of practice, cases and social media posts will encourage the reader to reflect on these questions and their own experience and thinking. Throughout this discussion, attempts have been made to open up questions of human action, thought and knowing, not only to be curious beings, to inquire and create, but to inquire and create ethically. It may be that questions of consciousness, language and inter-subjectivity are not accessible to introspection, but they have confounded philosophers and other citizens for many centuries and their complexity will continue to confuse and excite. For educators at all levels and within all cultures, these are challenging questions that cannot be denied if democratic, collective and honest learning is valued. For 30 years however, these questions have been denied by neoliberal capitalism, in favour of formal education that focuses on narrow economic and individual outcomes. These are questions that lie at the heart of social division and encourage different thinking by different groups of people.

At this stage, we need to review a number of key ideas encountered previously before bringing them together as cohesively as possible. I have tried to narrate my own thinking about philosophy and knowledge from the point of view of an educational practitioner. This applies to our understanding of philosophy itself. In the first book of this trilogy (Hooley, 2015), I argued that there is a fracture between what sociology describes and what philosophy as epistemology needs to do, or how humans act, to improve formal education for all agents. I called this gap between sociology and epistemology the 'practice interface' where teacher and other practitioners need to act every day. On this basis, philosophy for me must be 'naturalistic' in that it takes the problems that involve ordinary people and considers them in the context of other beliefs that people have, the moral, political, cultural, aesthetic, economic and religious beliefs that abound. For educators – and indeed for all people as knowledgeable citizens – this consideration is epistemological drawing on our past, present and future understandings at once. Philosophy is thoughtful social practice of living as we exist in the real and naturalistic world, a process of practice-theorising. How this might be done and in relation to current 'scientific' or accepted knowledge will be considered below. Althusser (2017, p. 192) added to this when he suggested that a philosopher 'fights in theory' and 'has to become a theorist through scientific practice and the practice of ideological and political struggle'. A broad context for philosophy must be taken into account, rather than isolated slices of reality. There are other related and difficult concepts that I have attempted to discuss in preceding chapters from my own perspective, such as those of pragmatism, epistemology and knowledge. Peirce was cited earlier as providing the original notion of pragmatism when he wrote that 'making our ideas clear' (to ourselves inwardly and to others outwardly) involves our whole conception of an object based on the total effects of that object both immediate and remote. Engaging objects or reality occurs through the 'social act' of Mead. For me, epistemology is about the nature of knowledge, how do we know what we know and questions of justification, belief, reality and what we call truth. These can be approached from logical analysis, but they are all issues of social action in the first instance. When questions of the nature and critique of personal, community and abstract knowledge are ignored in formal education, a situation of dogma is imposed. Finally, the practitioner needs to take into account two aspects of knowledge, that which is agreed as being a reasonable guide to reality arising from previous practice and that which involves perception of the connections between effects of the object that assess its application for particular events. Knowledge then, or as mentioned previously, 'to knowledge', is active, collaborative and subject to revision involving language and communication.

In coming to the notion of language as tool, Wittgenstein focused on language use or act, rather than initial meaning. If language can be thought of as an inner state of being for humans that connects with the external world then, similar to consciousness, language is a sensuous human disposition that generates perception and conception of various types depending on social and physical conditions. When snow falls, the young child does not need an extensive vocabulary to feel cold or to know the meaning of cold, or the wonder of softness and texture. Words will be added later to express these experiences and to communicate with others. Seen as a tool in this way, the inner state or mode of being will not separate causation, explanation, interpretation, feeling and the like, but will relate to the experience holistically. For the very young, it is difficult to understand exactly how perceptions produce specific thought when experience has not been had before, such as the first falling of snow. But the human organism will react based on the sum total of experience that has been accumulated to that time. In a similar way, humans will come to their own understanding of colour, sound and taste without words and without being taught by others. In conceptualising consciousness and language as inner states of being, the notion of 'language act' integrates other acts such as speech act and social act as humans pursue their quests for knowledge and understanding. We therefore conceive of humans as active, reflective beings, engaging with their inner and external worlds at all times from birth to death, coming to know and construct meaning both privately and publicly.

Philosophy of practice as totality

If we take the suggestion above of philosophy being thoughtful social practice of living as we exist in the real and naturalistic world, a process of practice-theorising, then it is necessary to envisage how this might occur, in social practice. This will be done below by combining four major ideas that have been developed in these pages as the narration and writing unfolded. A thought from James (1899/2015, p. 42, original emphasis) will assist as a starting point:

> It is astonishing how many mental operations we can explain when we have once grasped the principles of association. The great problem which association undertakes to solve is, *Why does just this particular field of consciousness, constituted in this particular way, now appear before my mind?* It may be a field of objects imagined, it may be of objects remembered, or of objects perceived, it may include an action resolved on.

I am not able to explain why *this particular field of consciousness* appears in my mind, but it seems to appear not from a vacuum, but from 'association' with other objects, thoughts and words that have been formed through practice. The notion that various associations, clumps, or societies of objects connect in the brain enabling comprehension of experience is a strong recommendation for respecting culture, beliefs and the like that constitute meaning and human 'wholeness'. We therefore refer to Figure 12.1 arising from discussion of bricolage.

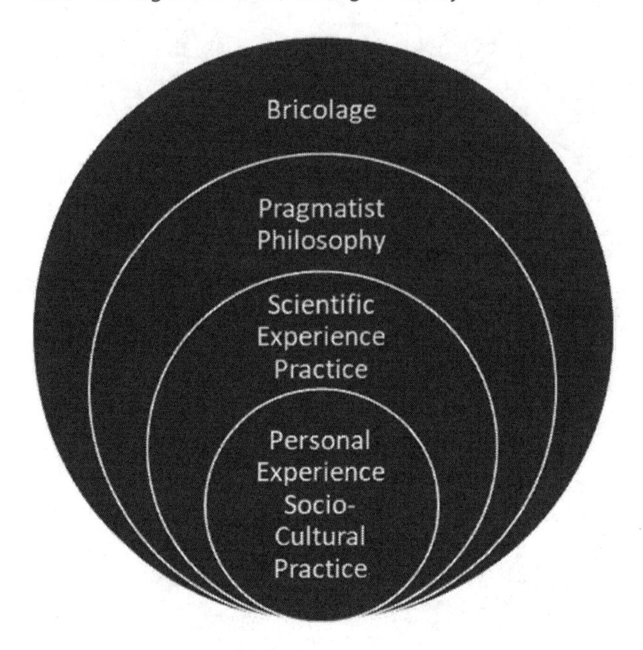

FIGURE 12.1 Philosophy of knowledge as practice

What I have termed the 'Pragmatist Research Shell' depicts four sets of acts and ideas in association with each other at all times. Living within the bricolage of socio-cultural and economic life, participants conduct inquiries into social problems that require resolution, bringing their formal and informal knowledges to bear (Bruce and Bloch, 2013). Notions of sociality and bricolage bear a strong resemblance to each other. Accordingly, working within the bricolage as bricoleur will necessarily change the disposition of researchers through their social acts and confrontation with all aspects of social life. For the purposes of this discussion, the concept of pragmatism has been used, whether this is expressed formally or informally by those individuals and groups concerned. Other paradigms can replace pragmatism if that is preferred. Thus, knowledge and learning are socially situation for all participants who are respected for the reality of the material conditions that they face and have faced; their experience is not denied. The 'Pragmatist Research Shell' can guide an investigation from tis design stage as well as monitor progress throughout. Next, the material conditions can be approximated and grouped in accord with Figure 12.2.

In this arrangement of the material conditions of existence known as knowledge, four broad groupings have been suggested that can be used in formal and informal settings such as research, community projects, family discussions, or educational interests. They are all components of a philosophical process defined above and relate to each of the shells in Figure 12.1. It is not necessary to distinguish between the human-derived domains of discipline and field, nor to include every single grouping such as mathematics, languages and the like, as this is a conceptual map that will alter in detail as each problem occurs. However it is submitted that the four groupings nominated cover the major associations of thought that participants refer to and infer from when working through their problems at hand. Aspects of 'working through' or knowledge production are noted in Figure 12.3.

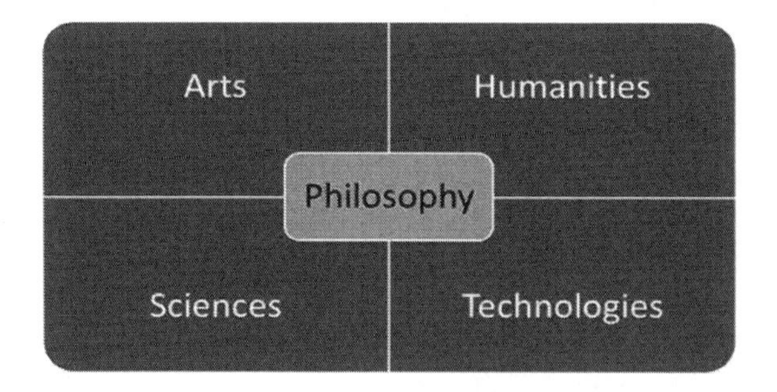

FIGURE 12.2 Philosophy of knowledge arrangement

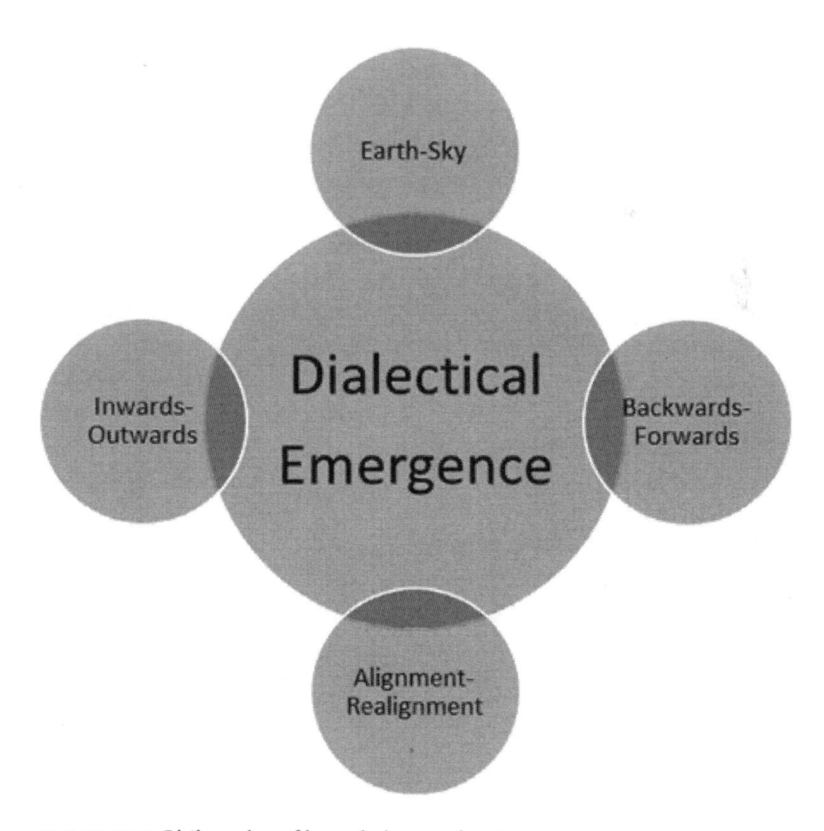

FIGURE 12.3 Philosophy of knowledge production

Four strategies in association are proposed as enabling the process of 'dialectical emergence' to take place, that is the process whereby new and novel ideas are constructed in relation to the problems that present and which occur through dialectical materialism, the motions and transformations of matter and energy. The human organism takes account of previous experience and its implications for ongoing action, its subjectivities and intersubjective understandings, its groundedness in experience and what this means for stable projections into the future and by so doing, enables alignment and realignment of knowledges and possibilities in its own and mutual interest. These processes are constant and continuous throughout existence and construct the organism's relation to the environment and how it appreciates the environment. Such a conceptualisation is shown in Figure 12.4.

Here we return to the notion of praxis at the centre of the construction of thought process where alignment and realignment of what we think involves again, in association and continuity, the constructs of imagination, description, explanation, theorising, change and reimagination. This is what ultimately brings forth the new. Intersubjective experience results from coming into contact with the situated acts and objects of existence from the position of the ideas, artefacts and acts that have occurred and from which the strategy of alignment and realignment is brought into play. A summary of the entire process indicated above is shown in Figure 12.5.

The above diagrams are not intended to constitute a mechanistic method for conducting an inquiry of any sort. Instead, they are conceptual and descriptive strategies, intending to proceed from the more general to the more specific and together, offer a philosophical and holistic view of knowledge formation, or in effect, human formation. Commenting on what they perceive as a tendency in American education towards methods that are said to work and therefore lead to the transmitting and verification of knowledge, Donald Macedo and Ana Freire in a foreword to Paulo Freire's book *Teachers as Cultural Workers* (Freire, 2005, p. x) write:

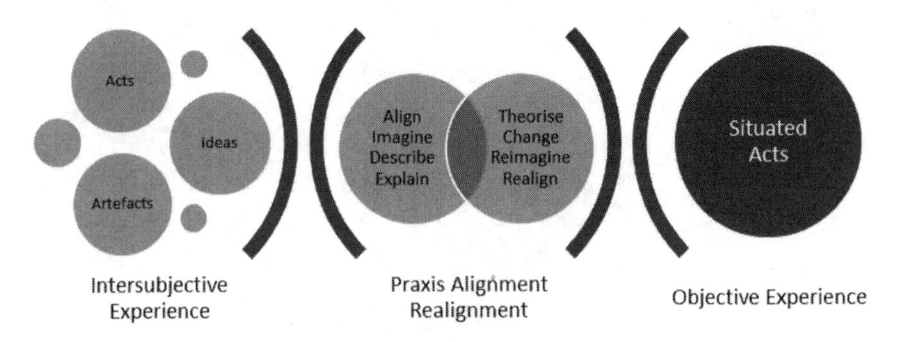

FIGURE 12.4 Philosophy of praxis alignment/realignment

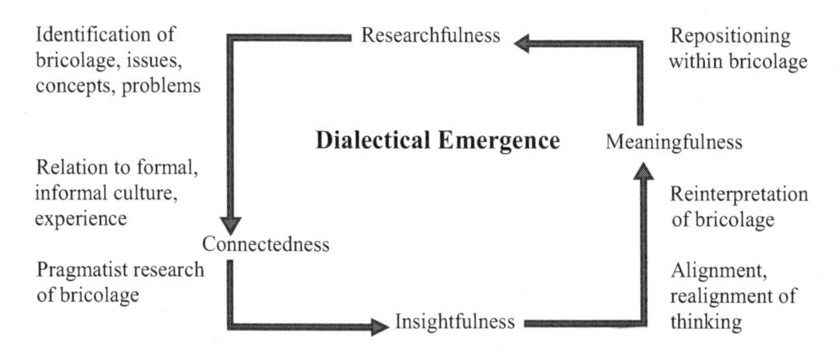

FIGURE 12.5 Representation of Pragmatist Research Shell

> This fetish for method works insidiously against the ability to adhere to Freire's own pronouncement against importing and exporting methodology. In a long conversation Paulo had with Donaldo Macedo about this issue, he said: 'Donaldo, I don't want to be imported or exported. It is impossible to export pedagogical practices without reinventing them. Please tell your fellow American educators not to import me. Ask them to recreate and rewrite my ideas.'

With this exclamation, Freire is encouraging educators not only to reconsider the question of method in terms of what they do, but in terms of what they think, of society, objects and of themselves. While there is place within an overall paradigm and methodology of inquiry for particular techniques, knowledge itself does not occur from predetermined, mechanical steps, but evolves from the situation at hand in collaboration with others through the adoption of imaginative strategies that connect with human consciousness and creativity. Children too must not have methods imposed but be able to rewrite their own ideas by being immersed in a portfolio of strategies that demonstrate their human capabilities as knowledgeable and anthropological agents of transformation.

Positioning educators philosophically

I now want to discuss one final example of the inherent weakness of neoliberal capitalism in relation to schooling and how the above epistemological strategies can provide democratic, high quality and defensible approaches in contrast. Overwhelming, current testing in schools is established on the wrong philosophical basis. It does not reflect the predominant ways by which humans learn and formulate knowledge. It distracts from the authentic process of knowing (Pratt, 2016). As Bruner (1979) commented, an educated person needs to know what knowledge is like, how it emerges from and integrates with other broader associative knowledge and how learning itself proceeds. This is personal learning of what is real for one rather than truth for the other. Testing knowledge (or more likely, preordained information) in a way that does not draw on personal culture and experience is like being coached in tennis and being tested on swimming; the results will be incomplete, inaccurate and distorted. As Hooley (2015, pp. 75–76) points out, it is traditional testing that is deficient, not the learning of children. He summarises:

- Knowledge in formal educational programs is often considered from a logical positivist perspective. This assumes a linear progression from knowledge that can be known accurately, taught, learned assessed and graded accurately. Experiential knowledge on the other hand, involves cycles of practice where knowledge is built, discussed, monitored, evaluated by consensus and ungraded as a guide to and support for further learning.
- Assessment that assumes one approach to teaching and learning imposes an ideological and epistemological view of knowledge that accepts power and control by those who assume to know. Under these circumstances, superficial learning in any class is restricted to a proportion of students and excludes those who prefer to learn in an interdisciplinary and inquiry manner, drawing upon aesthetic, scientific and communicative experience.
- Unfortunately, a recent report from the Grattan Institute in Australia (Goss and Sonnemann, 2016) ignores these issues and claims that a learning gap between students from different socio-cultural backgrounds is wide and increases throughout schooling. This claim is based on data from NAPLAN (2017), a national programme of mass testing in Australia, but is flawed in a number of respects:

- No philosophical or epistemological critique of mass testing regimes and what they purport to test and measure, highly inaccurately and with false assumptions.
- No critique of the NAPLAN scales; test scales are a social construction decided by a committee and could quite easily be different.
- No educational critique of the concept of student 'progress' or 'growth', entirely contained within predetermined limits and tested accordingly, i.e. growth is defined as test results between two points on a predetermined content scale.
- No epistemological discussion of knowledge, teaching and learning itself, but assumptions about teacher quality within current knowledge paradigms.
- Inadequate discussion of the need for more nuanced data and analysis to critique findings, e.g. correlations between social background or family income and test results are decidedly uncertain.

As incorrectly inferred by the Grattan report, there is no reason to suspect that children from certain socio-economic backgrounds are intellectually weak in comparison to children from other socio-economic backgrounds based on highly dubious test scores. Nor, that some students are 'disadvantaged' intellectually in comparison to some others. These assertions make no sense whatsoever, that vast numbers of children are 'disadvantaged' in their thinking, an evolutionary mistake in the DNA. As Ravitch (2013, p. 263) has explained, '. . . it is students, not teachers, who ultimately decide whether they want to learn more' and '. . . assumption that standardised tests are the best way to measure learning has never been established'. Standardised high stakes testing regimes like NAPLAN and PISA (Sellar and Lingard, 2014) are deliberately designed to measure narrow bands of predetermined information and skill that can be disconnected from the learner's personal, community and unorthodox experience (Rutkowski and Rutkowski, 2016). This view of what it means to be human is erroneous and totally misrepresents the capabilities of most of the student population. It may be an intentional ploy.

Connecting with lifeworld as the basis of social justice

Matthew Arnold wrote about society and education of the 1860s when Europe and England were attempting to cope with the Industrial Revolution and the new social forces that were being unleashed. Arnold was born in 1822, the eldest son of Thomas Arnold, the headmaster of Rugby School, and attended Oxford University. At Oxford, Arnold was influenced by Cardinal John Newman who wrote about the nature of universities and the concept of 'liberal education' for all of England. Matthew Arnold became Inspector of Schools in 1851and continued in that role until 2 years before his death in 1886. Being deeply religious, Arnold saw human 'perfection' as doing God's will during a time of social upheaval and demands being made by the emerging industrial working class. He proposed the notion of 'culture' as a stabilising influence during this time such that 'The pursuit of perfection, then, is the pursuit of sweetness and light'. He detailed this inspiration in the following way (Wilson, 1932/1984):

> I have been trying to show that culture is, or ought to be, the study and pursuit of perfection and that of perfection, as pursued by culture, beauty and intelligence, or, in other words, sweetness and light, are the main characters. But hitherto, I have been insisting chiefly on beauty, or sweetness, as a character of perfection. To complete rightly my design, it evidently remains to speak also of intelligence, or light, as a character of perfection.

Arnold did not see culture being imposed on the working class, but rather the purpose of culture was 'to make the best that has been thought and known in the world current everywhere'. In a famous statement, he proposed that 'This is the *social idea* and the men of culture are the true apostles of equality.' In considering these ideas 150 years later, we need to be able to have our own concept of culture, how it intersects philosophically with modern education and how associated programs might be implemented. This is particularly important in terms of assumptions made about public and private schooling and the social and cultural background of families. Clearly, there is debate about 'the best that has been thought and known' and the values, practices and meaning on which such debate is conducted. There is debate about beauty, intelligence, truth and goodness and how these virtues are understood and defined. The purpose of formal schooling is not agreed let alone the approaches by which children are encouraged to participate with knowledge and generate their own learning. It is possible for the school curriculum to divide knowledge into a number of areas such as arts, language, physical education, mathematics, sciences and humanities, with highly contested views regarding the nature of knowledge and how learning occurs. In other words, the notion of 'sweetness and light' is far from settled.

Mathematicians sometimes proclaim that the equation or proof is so beautiful that it must be true. This relates to the internal logic and coherence of the process followed as it is derived from agreed principles and postulates and how the outcome describes a particular aspect or relationship of the universe. The notion of truth applies to human acceptance of the accuracy of a statement, whereas the notion of beauty refers to the quality of an experience. Being able to follow the derivation of a symbolic statement that explains the properties of a pattern that appears to hold throughout the natural world, can be thought of as engaging the truth of the universe and of revealing the beauty of the universe. Humanity has not as yet agreed on the purpose of its own existence, but the practice of contemplation whereby such questions are considered and investigated over time and a closer understanding of reality and meaning emerges is of central concern. The concepts of truth and beauty are human concepts that belong to us all, regardless of social background, although background will influence our interpretation. The beauty of a rose will be apparent in the consciousness of the construction worker and of the aristocrat, although the former may be attracted to the layers of petals as sheets of materials, while the latter may be attracted to fragility and scent. The weeping willows of a Monet painting may be deeply evocative for the returned soldier as an expression of grief for war, but may be merely interesting as artistic technique for the bank manager. Or the reverse. Knowledge, learning and virtue as essential elements of culture need to be accessible for all children in school.

Experiencing the truth and beauty of a rose, painting, poem or equation within the broad conduct of social life constitutes key aspects of our humanity. It is the role of schools to create learning environments that enable this experience to flourish rather than to impose knowledge that is sterile and predetermined. It is to be expected that Arnold would not have a detailed understanding of knowledge and that he would assume that it was 'high' culture that should be available to everyone. However he stated that the best knowledge should be carried 'from one end of society to the other' and that knowledge should be divested 'of all that was harsh, uncouth, difficult, abstract, professional, exclusive; to humanise it, to make efficient outside the clique of the cultivated and learned'. One way of doing this is to adopt the approach of Charles Dodgson who, as Lewis Carroll, opened up mathematical ideas and thinking to millions of children through literature, words, puzzles and paradoxes. Writing at the same time as Arnold, Dodgson (who also attended Rugby and Oxford) drew upon the language, curiosity and inquisitiveness of the child in wanting to know. This is the moral approach to learning, where compassionate humans share their experience and investigate issues together, taking different pathways and reaching unexpected incomplete outcomes. Curriculum is a selection from all that is possible and there is little evidence to suggest that the imposed, pre-set curriculum involves children deeply in personal knowledge production and meaning.

Aristotle spoke of phronesis as the human disposition to live well and praxis as human action to put this disposition into effect. Given the lack of philosophical justification of schooling, we can combine Aristotle's notion of praxis as ethically-informed action for the public good and Arnold's notion of culture as knowledge that is the best that has been thought and known, as sweetness and light. Knowledge is not imposed as humans investigate their universe, but investigations can proceed within a framework of recognised understanding – although this needs to be negotiated with others: 'When *I* use a word, it means just what I choose it to mean, neither more nor less' (Humpty Dumpty). There is no need to assume that children will 'miss out' on essential knowledge if it is not mandated, especially as there is no defence of children 'missing out' under present arrangements that is not embracing, comprehending, or making intelligible presented knowledge. The praxis approach to personal knowledge establishes fecund environments where learning proceeds through social action, where ideas emerge from investigation of human interest and where outcomes of investigation provide open circumstances for further experiment and reflection. Praxis schools and classrooms will look quite different to the conservative teacher-directed model that has remained unaltered for many decades. Assessment of praxis knowledge and learning will also be different with traditional testing being abolished as epistemologically inappropriate and immoral.

Dialectics of knowing in social life

I hope that the discussion outlined in this book will assist readers in considering some of the great historical questions indicated at the beginning, those of 'How should we live?' 'What can we know?' and 'What does it mean to experience mind, to act, think, know and create ethically?' Education, in the sense described in this book, should be a major contributor to this endeavour, towards human satisfaction, justice and peace. It is the general concept and process by which we interact with the world and by so doing, come to deeper understandings about the realities of social and physical existence. Education is philosophical social practice. Kennedy (2017, p. 207), for example, in his discussion of 'Communal Philosophical Inquiry and *Schole*' describes school as interrupting the culture of state and economic production in terms of 'Platonic *theoria* – beholding or contemplating through the "eye of the mind" or *noesis*'. Institutionalised education therefore is recognised by all countries as being central to political and economic structures and international relationships, as well as being an indicator of national culture, harmony and dignity. Whether or not the ideas and practices that have been outlined in this book will provide more substantial means for considering these issues to some extent remains for readers to determine. However from my perspective and in relation to educational practice, writing and knowledge, some generalised remarks can be made about intent, organisation of the book and the process of writing and thinking, as follows:

1. Dialectical materialism and contradiction are processes that can be seen throughout the universe and in all aspects of social life. Involving the motion and transformation of matter and energy, the features of history and the components of personal and community interaction, dialectical materialism and contradiction are the dynamic forces that underpin change and progress. Unfortunately, we still live in the human era of war and aggression, but this will be transformed into consensus and peace in due course. Trends in education and knowledge can be traced throughout the centuries and indicate that progressive approaches involving aspects of dialectical materialism and contradiction are usually present, although often dominated by conservative ideologies. Substantial and enduring educational change will only occur when controlling economic and political systems are transformed into more democratic and citizen-oriented programs that recognise education as a process of becoming more human, more ethical and more compassionate beings.
2. Structuring the book in three parts around the ideas of looking backwards and forwards, looking inwards and outwards, and looking to the earth and sky, mirrors what has been suggested as the general manner by which humans organise their thinking and doing. This is speculative, but it accords with observation and allows for the integration of all aspects of our interactions with the world. As will be mentioned below, this organisation has brought forward a number of theorised ideas that will impact on the organisation of formal education and make schooling a more satisfying experience for many more students. In that regard, the organisation of schooling itself including curriculum could be structured around these three principles such that all students are able to draw upon their personal experience when considering all topics and locate their evolving understanding within the appropriate bricolage. This will require a degree of flexibility and indeed radicalism towards knowledge and learning that cannot exist within conservative paradigms.
3. Pragmatist theories and theorists have guided construction of the book and how ideas and practices have been engaged. At its heart, pragmatism encourages all humans to social action as the process by which we construct our understandings, values and ethics of what it means to be human. Dewey, Mead, Freire, Vygotsky and more recently Biesta have been the main theorists that have excited and challenged this work with their dedication, persistence and thoroughness being stimulating and inspirational. In contrast to the conservative viewpoint, these thinkers and writers attempt to take as many

factors as possible into account and reflect on issues in a critical and all-sided way. They place education at the centre of social life, not at the periphery. Kincheloe had a similar approach in his scholarship on critical pedagogy and the need to be generous and open-minded rather than narrow and dogmatic. What this means is that practitioners must be as well read as possible so that the work of others can be respected and constantly inform the practice-theorising of contemporary problems.

4. Not all realisations that are created throughout social or educational life are original and that has certainly been the case with the production of this book. However it is incredibly significant to realise that at a particular point in the accumulation of formal and informal experience, similar positions are reached to those of other and esteemed colleagues. In fact, that is the purpose of theory and theorising, to inform and guide further practice as problems continue to arise for resolution. To suddenly realise in one's mind that the views of Darwin seem reasonable, that the social act of Mead may be the starting point of cognition, that the struggle of Pankhurst for universal suffrage strengthened cultural life today, or that Hawking's thinking about the destruction of information falling into black holes shows the development of modern science, means that we move forward together. We are all attempting to work through our confusing relationship with the world of persons, objects and ideas, taking into account what others think, but doing our own thinking at the same time. Each sentence is original, each thought is distinctive, each act is significant. Children must be respected for doing exactly the same.
5. Writing is a significant aspect of the practice-theorising process. Major ideas that have been generated through the construction of the book were not known at the beginning, or were not realised to the same extent. Theorising such as 'to knowledge' and 'dialectical emergence', together with 'Pragmatist Research Shell' occurred through the process of reading, thinking and writing day-by-day, although of course, were stabilised on the structure of many years educational practice that has been documented and discussed. As the work continued, I thought that Chapter 12 would become a place where other generalisations would appear and that did happen. The five diagrams in this chapter make a coherent whole for educators to frame their thinking about research, knowledge and the organisation of educational practice. Adoption of this framework around the world will contest the domination of neoliberal capitalist exploitation of education at all levels. This did not occur to me until late in the writing and was not finalised until Chapter 12 itself was in process. Attempting to write in a personal narrative style, very much in accord with the 'Pragmatist Research Shell' concept, enabled a more fluid progression of idea association, rather than step-by-step procedures that restrict thinking and imagination.
6. Progressive approaches to education and knowledge proceed as a totality, rather than the disconnected features of conservative ideology. The discussion of mass testing above shows this clearly. Dialectical materialism and pragmatism recognise that everything is connected to everything else, that this is the way

humans engage the world and that to deny such connections diminishes humanity. This is the significance of the bricolage in research and knowledge and how the bricolage enables continuing participation by different groups such as Indigenous, feminist and ethnic and the incorporation of different culture and experience. Refusal to conceive of life as a totality, of education as a totality and a refusal to take these issues into account results in exclusion from social practice and alienation from education whether formal or informal. When epistemology, pedagogy, curriculum design and assessment procedures lack progressive and critical philosophical coherence, staff, students and families are adrift without a map and compass that makes the educational journey frustrating and often, meaningless.

7. Learning and knowledge arise from social practice, they do not drop from the sky or are the preserve of a few. That being so, acknowledgement is given to all colleagues who have taken this journey with me. This involves the theorists mentioned above, but it includes most importantly those who have worked side-by-side each day over many years. In the fine tradition of dialectics and contradiction, there is not always agreement, but when colleagues come together with mutual interest to improve educational practice, the outcome is always beneficial, whether local or global, present or historical. Attempting to establish a paradigm of respectful practice and praxis is a significant contribution to social and educational understanding that is a credit to all concerned. Accordingly, *Dialectics of Knowing in Education* has resulted as a manifesto for practitioners to change the world for the betterment of humanity.

It is unlikely that dialectical knowing in education as described throughout this narrative will have a marked effect on social division within neoliberal society dominated by capitalist imperatives. It will however have a discernible impact on education systems making them less divisive, more democratic and more equitable (Mason, 2017). Exactly how to do this will depend on the socio-political and cultural conditions that direct and regulate the structures and classifications that transpire. For educators who live and work within such systems, individual and collegial decisions will need to be made with care and flexibility enabling the complex web of connections across the bricolage to inform judicious judgement and successive social acts. Learning circles or small groups of participants need to invite all staff, student and researcher members alike into their personal understandings with trust and respect so that intimate details of practice and knowledge can be shared and critiqued; sectarianism and triumphalism have no place in democratic learning. Research or general projects in all curriculum domains no matter how organised should be negotiated on the basis of participant interest, with provisional or reasonably definite outcomes subject to reflective evaluation and appraisal. Assessment of all project conclusions apply not only to project questions themselves, but to the knowledge and ethical impact that participation has had on all investigators. Consequently, how we arrange our approach towards knowledge in all aspects of our social lives strongly influences our capabilities as humans to live and love democratically, to accept our responsibility for communal satisfaction and well-being and to stand alongside others in constructing a more noble and generous humanity for all. This is our historical legacy.

With best collegial wishes for peace and justice

Neil

Case 12. La Sagrada Familia

My friend Leo had just returned from holidays in Spain. He was enthusiastic about his visit to Casa Mila, an apartment building in Barcelona apparently, I was told, designed around the innovative use of curves, undulations, twists and turns and iron balconies. As a former science teacher, he was fascinated by the incorporation of various arches as practical and beautiful features of architecture. No doubt we had both mentioned catenary and parabolic curves in our classes from time to time, probably not with great authority, but with attempts to relate mathematics and nature. I did recall holding a small chain at both ends and noting the shape it formed as it dropped in the middle. 'I know you respect my catholic background,' said Leo with that ironic smile of his, 'but it's well worth a look from a historical, cultural and architectural point of view.' Before I could respond with a biting witticism, he went on: 'But if you really want to be amazed, you should visit the Sagrada Familia.' Now I haven't travelled a great deal in Europe and I certainly do not have interest in touring yet another cathedral, but I had heard of this famous building still incomplete with its unique sweeping spires and arches outside and many twisting columns inside. Perhaps I had seen a documentary in the distant past about the architect Gaudi. By this stage Leo was in full stride, explaining that Gaudi had used his knowledge from observing nature to construct the internal load-bearing columns through the twisting of various geometric forms. 'Well, I admit it would be interesting as a mathematics and physics excursion,' I managed to interrupt, 'but I haven't been to the Acropolis yet, to walk in the footsteps of Aristotle, Plato and the others.' Leo knew of my interest in Greek philosophy and my argument that this was a huge gap in the formal schooling for all students regardless of their backgrounds. 'Sure, that would be great as well, but the influence of the philosophers, artists, artisans and architects can be seen across all of Europe, including the churches. Goya, Caravaggio, Rapfael, they are all there.' Anticipating what I was about to say, he quickly added, 'And yes, even Michelangelo!' As I grappled with Leo's enthusiasm, I could not help but feel a little worried that I was, after all these years, finding his argument more convincing. Perhaps I was being too narrow-minded in not recognising the diverse places where human development could be traced and, after all, as a progressive educator, I had an obligation to look at the totality of evidence, of human experience. But, as an atheist, it had always been difficult for me to connect religion and science as contradictory aspects of knowledge. Could each really be found in the other?

'You know, Gaudi must have had wonderful imagination and creativity and you are always going on about how you admire the thinking and logic of the Greeks and others throughout history, even if we now know that they got some things wrong,' insisted Leo, pressing home the point. 'He made models, rather than sketches I believe, that's amazing, he must have thought in three dimensions, wow.' It was one of those occasions when you just feel, deep inside, that a new idea, or an old idea long rejected, was making more sense, bubbling to the surface. Perhaps, for some reason, I was a different person today than before, or some extra pieces of the jigsaw had been added, or chatting informally with a respected friend of different perspective had opened my eyes to fresh possibility. Whatever the case, I gathered it was time to take my own advice often provided to students, that we should do our own independent thinking regarding the totality of a situation and not act superficially on dogma or doctrine. 'Hey Leo, after my exciting mathematical exploration of La Sagrada Familia and Casa Mila, could I visit sites of the Spanish Civil War, that would be something?'

Knowledge Exemplar 3

I tend to agree with Bernstein (1971/1999, p. 80) when he claimed that 'the dialectic that can take place between Marx and Dewey is the political dialectic of our time'. Perhaps Hegel should be included here as well. Although this was written nearly 50 years ago during the Welfare State period, it remains significant and relevant today with neoliberalism in a dominant position around the world. Since Bernstein's comment, a more nuanced view can be recast as the current dialectic between the 'big ideas' and theories of economics, evolution, relativity and pragmatism at the beginning of the twentieth century and the 'big ideas' and theories of quantum mechanics, large data, artificial intelligence and biotechnology and genetics of the twenty-first century. In some respects, the political dialectic involves the revolutionary outcomes of Marx where one social and economic system is replaced by another, compared with the liberal changes of Dewey that, on the surface, seek to improve but take place within capitalist arrangements. However it was suggested in earlier pages that if the pragmatist approach envisaged by Peirce was fully implemented and our knowledge and understandings were formed through our conception of the total effects of an object, then this would lay the basis for revolutionary social change, ultimately. That is, a more profound and more accurate appreciation of the world and the place of ethical humanity within it, than is the case under neoliberalism. Knowledge Exemplar 3 (Table 12.1) below is an attempt at outlining how this intersubjective process of interacting with the objects of the world could proceed informally, as individuals and groups go about their daily work, school and home activities and formally, when we are called upon to construct and critique our ideas and beliefs through systematic investigation. Each approach falls within the 'Pragmatist Research Shell' of social action described previously. Pragmatism is not a programme for political revolution. It may have struggled to ask many questions regarding cultural and economic life

TABLE 12.1 Knowledge Exemplar 3: philosophy of practice

Philosophy of Practice	*Indicator 1* *Informal Cultural*	*Indicator 2* *Informal Community*	*Indicator 3* *Formal Professional*	*Indicator 4* *Formal Theorising*
Pragmatist Research Shell	Relating new events to individual experience	Drawing on family, group links with experience	Connecting features of bricolage	Planning critical investigation within bricolage
Knowledge arrangement	Integrating thoughts from different content domains	Prioritising across different content domains of community action	Specifying relevant knowledge domains	Recognising main features of knowledge base
Knowledge production	Making sense from personal perspectives	Giving preference to community understandings	Making value judgements about perspectives	Emphasising and justifying main viewpoints
Knowledge alignment, realignment	Creating novelty when known and unknown associate	Evaluating new possibilities for interest of community	Assessing outcomes relative to questions	Interpreting meaning with ethical intent and personal impact

throughout the war-torn twentieth century and may not have identified itself within a critical bricolage of national and international power and intrigue. But this situation can be corrected by radical pragmatists and bricoleurs acting together to transform, for example, the field of education into its opposite for families and their children everywhere. In the end, education that does not seek to liberate all, degrades and dishonours all.

POEM

Story lines

Thoughts caressed by a gentle breeze
positioned in the luxuriance of green
history observed by patient trees
eternal in my fragile life,
light transforms the finch's wing
and time encircles evocative ideas.

Harmony descends across the land
drawing in through sheltered arms
embracing all in perspective calm
confidence in divulging doubt,
buoyant trout in the crystal stream
and gravity pulls at golden leaves.

Listen to the stories of the land
crafted by the endless tides
suggesting what might endure
against the convulsions of change,
seeds germinate in the sunburnt soil
and resonate with the fall of rain.

Unhurried journeys open closed eyes
paths once hidden by noise and greed
expose the possibility of meaning
connecting what exists in depth of mind,
anticipation swells across the landscape
and wild flowers converse with all.

REFERENCES

ACARA. (2017). Australian Curriculum, Assessment and Reporting Authority, accessed at www.acara.edu.au/, 1 December 2017.

ACER. (2017). *PIRLS 2016 Highlights from Australia's perspective, selected findings from the full report 'Reporting Australia's Results PIRLS 2016'*, Australian Council for Educational Research: Melbourne, accessed at https://research.acer.edu.au/cgi/viewcontent.cgi?article=1001&context=pirls, 15 December 2017.

ACTU. (2018). Australian Council of Trade Unions, accessed at www.actu.org.au/, 7 January 2018.

AEC. (2015). *National STEM School Education Strategy: A Comprehensive Plan for Science, Technology, Engineering and Mathematics Education in Australia*, Canberra: Australian Education Council.

Allison, P. and Pomeroy, E. (2000). How shall we 'know'? Epistemological concerns in research in experiential education. *The Journal of Experiential Education*, 23, 91–98.

Althusser, L. (2017). *Philosophy for Non-Philosophers*, translated and edited by Goshgarian, G. M., London: Bloomsbury.

Angelou, M. (2018). Maya Angelou, *Biography*, accessed at www.biography.com/people/maya-angelou-9185388, 3 January 2018.

Arendt, H. (1958). *The Human Condition*, Second Edition, Chicago, London: University of Chicago Press.

Arnold, J., Edwards, T., Hooley, N. and Williams, J. (2012). Conceptualising teacher education and research as 'critical praxis'. *Critical Studies in Education*, 53(3), 281–295.

Aronson, B. and Laughter, J. (2016). The theory and practice of culturally relevant education: A synthesis of research across content areas. *Review of Educational Research*, 86(1), 163–206.

ASIC. (2017). 'The algorithm ate my homework' is no excuse; ASIC, *Financial Review*, accessed at www.afr.com/business/banking-and-finance/financial-services/the-algorithm-ate-my-homework-is-no-excuse-asic-20170910-gyed3j, 16 December 2017.

Australian Women's Register. (2018). Oodgeroo Noonuccal 1920–1993, The Australian Women's Register, accessed at www.womenaustralia.info/biogs/IMP0082b.htm, 9 January 2018.

Ball, S. J. and Olmedo, A. (2013). Care of the self, resistance and subjectivity under neoliberal governmentalities. *Critical Studies in Education*, 54(1), 85–96.

Barnes, J. and Kenny, A. (2014). *Aristotle's Ethics: Writing From The Complete Works*, Princeton, Oxford: Princeton University Press.

Bernstein, R. J. (1971/1999). *Praxis and Action: Contemporary Philosophies of Human Activity*, Philadelphia: University of Pennsylvania Press.

Biesta, G. J. J. (1998). Mead, intersubjectivity, and education: The early writings, *Studies in Philosophy and Education* 17, 73–99.

Biesta, G. J. J. (1999). Redefining the subject, redefining the social, reconsidering education: George Herbert Mead's course on philosophy of education at the University of Chicago, *Educational Theory*, 49(4), 475–492.

Biesta, G. J. J. (2006). *Beyond Learning: Democratic Education for a Human Future*, Boulder, London: Paradigm Publishers.

Biesta, G. J. J. (2006a). 'Of all affairs, communication is the most wonderful'. The communicative turn in Dewey's Democracy and Education, in Hansen, D. Y. (Ed.) *John Dewey and our Educational Prospect: A critical Engagement with Dewey's Democracy and Education*, Albany: State University of New York, pp. 23–38.

Biesta, G. J. J. (2010). *Good Education in an Age of Measurement*, Boulder: Paradigm Publishers.

Biesta, G. J. J. (2013). *The Beautiful Risk of Education*, Boulder: Paradigm Publishers.

Biesta, G. J. J. (2017). *Art Education 'After' Joseph Beuys*, Arnhem: ArtEZ Press.

Biesta, G. J. J. (2017a). *The Rediscovery of Teaching*, New York, London: Routledge.

Biesta, G. J. J. and Burbules, N. C. (2003). *Pragmatism and Educational Research*, New York: Rowman & Littlefield.

Biesta, G. J. J. and Trohler, D. (Eds). (2008). *The Philosophy of Education: George Herbert Mead*, Boulder, London: Paradigm Publishers.

Bishop, R., Ladwig, J. and Berryman, M. (2014). The centrality of relationships for pedagogy: The Whanaungatanga thesis. *American Educational Research Journal*, 51(1), 184–214.

Blackmore, J. (2013). Within/against: Feminist theory as praxis in higher education research, in *Theory and Method in Higher Education Research*, Emerald Group Publishing: London, pp. 175–198, accessed at http://dro.deakin.edu.au/eserv/DU:30061591/blackmore-withinagainst-post-2013.pdf, 9 January 2018.

Boaler, J. (2016). *Mathematical Mindsets: Unleashing Students' Potential through Creative Math, Inspiring Messages and Innovative Teaching*, San Francisco: Jossey-Bass.

Bourdieu, P. (1977). *Outline of a Theory of Practice*, Cambridge: Cambridge University Press.

Bourdieu. P. (1984). *Distinction: A Social Critique of the Judgement of Taste*, Oxford: Routledge.

Boyte, H. C. (2017). John Dewey and citizen politics: How democracy can survive artificial intelligence and the credo of efficiency, *Education & Culture*, 33(2), 13–47.

Brownfoot, J. N. (1983). Goldstein, Vida Jane 1869–1949, *Australian Dictionary of Biography*, accessed at http://adb.anu.edu.au/biography/goldstein-vida-jane-6418, 11 January 2018.

Bruce, B. C. and Bloch, N. (2013). Pragmatism and community inquiry: A case study of community-based learning, *Education & Culture*, 29(1), 27–45.

Bruner, J. (1979). *On Knowing: Essays for the Left Hand*, Cambridge, MA: Cambridge University Press.

Burgh, G., Field, T. and Freakley, M. (2006). *Ethics and the Community of Inquiry: Education for Deliberative Democracy*, South Melbourne: Thomson Social Sciences Press.

Byers, W. (2007). *How Mathematicians Think: Using Ambiguity, Contradiction and Paradox to Create Mathematics*, New Jersey: Princeton University Press.

Cambourne, B. (2014). The seven messages of highly effective reading teachers, *The Conversation*, accessed at https://theconversation.com/the-seven-messages-of-highly-effective-reading-teachers-24777, 5 February 2018.

Capstone Writer. (2009). *Producing the Capstone Project*, Morrisville: Lulu.com.

Carr, W. and Kemmis, S. (1986). *Becoming Critical: Education, Knowledge and Action Research*, Geelong: Deakin University Press.

Chalmers, D. J. (1996). *The Conscious Mind: In Search of a Fundamental Theory*, New York, Oxford: Oxford University Press.

Cherednichenko, B. and Kruger, T. (2005/2006). Social justice and teacher education. *International Journal of Learning*, 12(7), 321–332.

Chesky, N. (2014). Policy and praxis: An ontological study of U.S. mathematics discourses, *Philosophy of Mathematics Education*, 28, accessed at http://socialsciences.exeter.ac.uk/education/research/centres/stem/publications/pmej/pome28/index.html, 25 January 2018.

Connell, R. (2010). Kartini's children: On the need for thinking gender and education on a world scale. *Gender and Education*, 22(6), 603–615.

Connell, R. (2013). The neoliberal cascade and education: An essay on the market agenda and its consequences. *Critical Studies in Education*, 54(2), 99–112.

Connell, R. W. (1993). *Schools and Social Justice*, Philadelphia: Temple University Press.

Craft, A. and Jeffrey, B. (Eds). (2008). Special Issue: Creativity and performativity in teaching and learning: Tensions, dilemmas, constraints, accommodations and synthesis, *British Educational Research Journal*, 34(5).

Creswell, J. (2014). *Research Design: Qualitative, Quantitative and Mixed Methods Approaches*, Fourth Edition, Los Angeles, London, New Delhi, Singapore, Washington DC: SAGE Publications.

Csikszentmihalyi, M. (1996). *Creativity: Flow and the Psychology of Discovery and Invention*, New York: HarperCollins Publishers.

Cuban, L. (2016). Educational researchers, AERA presidents and reforming the practice of schooling, 1916–2016. *Educational Researcher*, 45(2), 134–141.

Dakich, E., Watt, T. and Hooley, N. (2016). Reconciling mixed methods approaches with a community narrative model for educational research involving Aboriginal and Torres Strait Islander families. *Review of Education, Pedagogy, and Cultural Studies*, 38(4), 360–380, DOI: 10.1080/10714413.2016.1203683.

Darder, A. (2015). *Freire and Education*, New York, London: Routledge.

Darder, A., Mayo, P. and Paraskeva, J. (Eds). (2016). *International Critical Pedagogy Reader*, New York, Oxford: Routledge, 1–14.

Darling-Hammond, L. and Richardson, N. (2009). Research review/teacher learning: What matters? *How Teachers Learn*, 66(5), 46–53.

Davies, P. (2008). *The Goldilocks Enigma: Why is the Universe Just Right for Life?* Boston, New York: Mariner Books.

Davies, P. (2014). Universe from bits, in Davies, P. and Gregersen, N. H. (Eds) *Information and the Nature of Reality*, Cambridge: Cambridge University Press, 83–117.

Denzin, N. K, and Lincoln, Y. S. (Eds). (2000/2017). *The SAGE Handbook of Qualitative Research*, Fifth revised edition, Thousand Oaks: Sage Publications.

Dewey, J. (1897). My Pedagogic Creed, *School Journal*, 54, 77–80, accessed at http://dewey.pragmatism.org/creed.htm, 23 October 2017.

Dewey, J. (1910/1997). *How We Think*, Mineola: Dover Publications.

Dewey, J. (1916). *Democracy and Education*, London: Collier Macmillan.

Dewey, J. (1938). Logic: The theory of inquiry, in Boydson, J. (Ed.) *The Later Works (1925–1953) Volume 12*, Carbondale: Southern Illinois University Press.

Dewey, J. (1958). *Experience and Nature*, New York: Dover Publications.

Dewey, J. (1963). *Experience and Education*, New York: Macmillan.

Dreyfus, H. L. (1972/1992). *What Computers Still Can't Do: A Critique of Artificial Reason*, Cambridge, MA, London: The MIT Press.

Durst, A. (2010). *Women Educators in the Progressive Era: The Women Behind Dewey's Laboratory School*, New York: Palgrave Macmillan.

Dweck, C. (2012). *Mindset: How You Can Fulfill Your Potential*, New York: Constable and Robinson.

Eagleton, T. (2016). *Materialism*, New Haven, London: Yale University Press.

Eckersley, B., Davies, M., Arnold, J., Edwards, T., Hooley, N., Williams, J. and Taylor, S. (2011). *Vision Unlimited: Inspiring participant knowledge in schools. Researching Site-Based Pre-Service Teacher Education*, Melbourne: Victoria University.

Ek, A. and Latta, M. A. M. (2013). Preparing to teach: Redeeming the potentialities of the present through 'conversations of practice'. *Education & Culture*, 29(1), 84–104.

Engels, F. (1976). *Ludwig Feuerbach and the End of Classical German Philosophy*, Peking: Foreign Languages Press.

Ernest, P. (1998). *Social Constructivism as a Philosophy of Mathematics*, Albany: State University of New York.

Ernest, P. (2004). What is the philosophy of mathematics education? *Philosophy of Mathematics Education Journal*, 19, accessed at http://people.exeter.ac.uk/PErnest/pome18/contents.htm, 25 January 2018.

Fanon, F. (1961/2005). *The Wretched of the Earth*, New York: Gove Press.

Feenberg, A. (2014). *The Philosophy of Praxis: Marx, Lukacs and the Frankfurt School*, London, New York: Verso.

Feyerabend, P. (1975). *Against Method: Outline of an Anarchistic Theory of Knowledge*, London: Humanities Press.

Fielding, M. (2006). Leadership, radical student engagement and the necessity of person-centred education. *International Journal of Leadership in Education*, 9(4), 299–313.

Fraser, N. (2009). Social justice in the age of identity politics: Redistribution, recognition and participation, in Henderson, G. and Waterstone, M. (Eds) *Geographic Thought: A Praxis Perspective*, Oxford, New York: Routledge, 72–90.

Freire, P. (1970/2012). *Pedagogy of the Oppressed: 30th Anniversary Edition*, New York, London, New Delhi, Sydney: Bloomsbury Press.

Freire, P. (1972). *Cultural Action for Freedom*, Harmondsworth: Penguin Books.

Freire, P. (1972). *Pedagogy of the Oppressed*, Harmondsworth: Penguin Books.

Freire, P. (1992). *Pedagogy of Hope: Reliving Pedagogy of the Oppressed*, London and New York: Continuum.

Freire, P. (1993). *Pedagogy of the City*, New York: Continuum.

Freire, P. (2005). *Teachers as Cultural Workers: Letters to Those Who Dare Teach*, United States of America: Westview Press.

Fromm, E. (1956/2006). *The Art of Loving*, New York, London, Toronto, Sydney: Harper Perennial.

Gardner, H. (2006). *Five Minds for the Future*, Boston: Harvard Business School Press.

Gee, J. P. and Hayes, E. R. 2011. *Language and Learning in the Digital Age*, New York: Routledge.

Giddens, A. (1984). *The Constitution of Society: Outline of the Theory of Structuration*, Cambridge: Polity Press.

Giddens, A. (1993). *New Rules of Sociological Method: A Positive Critique of Interpretive Sociologies*, Cambridge: Polity Press.

Gore, J. (2017). Reconciling educational research traditions, *The Australian Educational Researcher*, 44(4–5), 357–372.

Goss, P. and Sonnemann, J. (2016). *Widening Gaps: What NAPLAN Tells Us about Student Progress*, Melbourne: Grattan Institute.

Grant, C. A. (2012). Cultivating flourishing lives: A robust social justice vision of education. *American Educational Research Journal*, 49(5), 910–934.

Greenfield, S. (2016). *A Day in the Life of the Brain: The Neuroscience of Consciousness from Dawn till Dusk*, Harmondsworth: Penguin Random House UK.

Hansen, D. T. (Ed.). (2006). *John Dewey and Our Educational Prospect: A Critical Engagement with Dewey's Democracy and Education*, Albany: State University of New York.

Hansen, D. T. (2014), Cosmopolitanism as cultural creativity: New modes of educational practice in globalizing times. *Curriculum Inquiry*, 44(1), 1–14, doi:10.1111/curi.12039.

Harding, S. (1986). *The Science Question in Feminism*, Ithaca, London: Cornell University Press.

Harrison, T. and Bawden, M. (2016). Teaching character through subjects. *Research Intelligence*, 130, 14–16.

Hartsock, N. (1998). Marxist feminist dialectics for the 21st century. *Science & Society*, 62(3), 400–413.

Hastings, M. (2012). The rise of the killer drones: How America goes to war in secret, *Rolling Stone*, accessed at www.rollingstone.com/politics/news/the-rise-of-the-killer-drones-how-america-goes-to-war-in-secret-20120416, 18 November 2017.

Hauhart, R. C. and Grahe, J. E. (2015). *Designing and Teaching Undergraduate Capstone Courses*, San Francisco: Jossey-Bass.

Hegel, G. W. F. (1977). *Hegel's Phenomenology of Spirit*, translated by Miller, A. V. and Foreword by Findlay, J. N., Oxford, New York, Toronto, Melbourne: Oxford University Press.

Heidegger, M. (1962/2008). *Being and Time*, translated by Macquarrie, J. and Robinson, E., New York: HarperPerennial ModernThought.

Hickman, L. A. (1992). *John Dewey's Pragmatic Technology*, Bloomington, Indianapolis: Indiana University Press.

Hiddleston, J. (2009). *Understanding Postcolonialism*, Ultimo New South Wales: Acumen Publishing.

Hodkinson, P. and Macleod, F. (2010). Contrasting concepts of learning and contrasting research methodologies: affinities and bias. *British Educational Research Journal*, 36(2), 173–189.

Hooley, N. (2010). *Narrative Life: Democratic Curriculum and Indigenous Learning*, Dordrecht, Heidelberg, London, New York: Springer.

Hooley, N. (2015). *Learning at the Practice Interface: Reconstructing Dialogue for Progressive Educational Change*, London, New York: Routledge.

Hooley, N. (2018). *Radical Schooling for Democracy: Engaging Philosophy of Education for the Public Good*, London, New York: Routledge.

Hopkins, N. (2014). The democratic curriculum: Concept and practice. *Journal of Philosophy of Education*, 48(3), 416–427.

Hostetler, K. D. (2016). Beyond reflection: Perception, virtue and teacher knowledge. *Educational Philosophy and Theory*, 48(2), 179–190.

Ings, W. (2015). The authored voice: Emerging approaches to exegesis design in creative practice PhDs. *Educational Philosophy and Theory*, 47(12), 1277–1290.

James, W. (1899/2015). *Talks to Teachers on Psychology and to Students on Some of Life's Ideals*, New York: Dover Publications.

Jeffries, S. (2016). *Grand Hotel Abyss: The Lives of the Frankfurt School*, London, New York: Verso.

Joas, H. (1997). *G. H. Mead: A Contemporary Re-examination of His Thought*, Cambridge, MA: MIT Press.

Kaag, J. (2011). *Idealism, Pragmatism and Feminism: The Philosophy of Ella Lynam Cabot*, New York: Lexington Books.

Kaptelinin, V. (2005). The object of activity: Making sense of the sense-maker. *Mind, Culture and Activity*, 12(1), 4–18.

Kartini. (2018). *Kartini, The Complete Writings 1898–1904*, edited and translated by Cote, J., accessed at http://publishing.monash.edu/books/kartini-9781922235107.html, 3 January 2018.

Keane, W. (2016). *Ethical Life: Its Natural and Social Histories*, Princeton: Princeton University Press.

Keddie, A. (2012). Schooling and social justice through the lenses of Nancy Fraser. *Critical Studies in Education*, 53(3), 263–279.

Kemmis, S. (2010). What is to be done? The place of action research. *Educational Action Research*, 18(4), 417–427.

Kemmis, S. (2012). Researching educational praxis: Spectator and participant perspectives. *British Educational Research Journal*, 38(6), 885–905.

Kemmis, S. and McTaggart, R. (1988). *The Action Research Planner*, Waurn Ponds: Deakin University Press.

Kemmis, S. and Smith, T. J. (2008). Praxis and praxis development, in Kemmis, S. and Smith, T. J. (Eds) *Enabling Praxis: Challenges for Education*, Rotterdam: Sense Publishers, 3–13.

Kennedy, D. (2017). An archetypal phenomenology of skhole. *Educational Theory*, 67(3), 273–290.

Kincheloe, J. L. (2005). On to the next level: Continuing the conceptualisation of the bricolage, *Qualitative Inquiry*, 11(3), 323–350.

Kincheloe, J. L. (2011). Describing the bricolage: Conceptualising a new rigor in qualitative research, in Hayes, K., Steinberg, S. R. and Tobin, K. (Eds) *Key Works in Critical Pedagogy*, Rotterdam: Sense Publishers, 177–189.

Kinsella, E. A. and Pitman, A. (Eds). (2012). *Phronesis as Professional Knowledge: Practical Wisdom in the Professions*, Rotterdam, Boston, Taipei: Sense Publishers.

Kohli, W. R. and Burbules, N. C. (2013). *Feminists and Educational Research*, Lanham, New York, Toronto, Plymouth: Rowman and Littlefield Education.

Kovach, M. (2009). *Indigenous Methodologies: Characteristics, Conversations and Contexts*, Toronto: University of Toronto Press.

Krashen, S. (2014). Stephen Krashen to LA School Board: Invest in libraries, *School Library Journal*, accessed at www.slj.com/2014/07/literacy/stephen-krashen-to-la-school-board-invest-in-libraries/, 5 February 2018.

Kuhn, T. S. (1962/2012). *The Structure of Scientific Revolutions*, 50th Anniversary Edition, Chicago: Chicago University Press.

Kuhn, T. (1977). *The Essential Tension: Selected Studies in Scientific Tradition and Change*, Chicago: University of Chicago Press.

Kutay, C., Mooney, J., Riley, L. and Howard-Wagner, D. 2012. Experiencing Indigenous knowledge online as a community narrative. *The Australian Journal of Indigenous Education*, 41(1), 47–59.

Lave, J. and Wenger. E. (1991). *Situated Learning: Legitimate Peripheral Participation*, Cambridge: Cambridge University Press.

Lawy, R., Biesta, G., McDonnell, J., Lawy, H. and Reeves, H. (2010). The art of democracy: Young people's democratic learning in gallery contexts. *British Educational Research Journal*, 36(3), 351–365.

Lenin, V. I. (1976). *One Step Forward, Two Steps Back: The Crisis in Our Party*, Peking: Foreign Languages Press.

Levi-Strauss, C. (1968). *The Savage Mind*, Chicago: The University of Chicago Press.

Lewin, K. and Lewin.G. W. (Ed.). (1948). *Resolving Social Conflicts: Selected Papers on Group Dynamics (1935–1946)*. New York: Harper and Brothers.

Loomba, A. (1998). *Colonialism/Postcolonialism*, London, New York: Routledge.

Lukacs, G. (1922/1968). *History and Class Consciousness: Studies in Marxist Dialectics*, Cambridge, MA: The MIT Press.

Malala. (2018). Malala Fund, accessed at www.malala.org/malalas-story/, 5 January.

Mao, T-T. (1968). On Practice, in *Four Essays on Philosophy*, Peking: Foreign Languages Press, 1–22.

Marx, K. (1932/1959). *Economic and Philosophic Manuscripts of 1844*, Moscow: Progress Publishers.

Marx, K. (1954/1974). *Capital: A Critical Analysis of Capitalist Production, Volume 1*, Moscow: Progress Publishers.

Marx, K. and Engels, F. (1888/1969). *The Communist Manifesto*, with an Introduction by Taylor, A. J. P., Harmondsworth: Penguin Books.

Mason, L. E. (2017). The significance of Dewey's *Democracy and Education* for 21st century education. *Education & Culture*, 33(1), 41–58.

McTaggart, R. (1991). *Action Research: A Short Modern History*, Waurn Ponds: Deakin University Press.

Mead, G. H. (1908). The philosophical basis for ethics. *International Journal of Ethics*, 18, 311–323.

Mead, G. H. (1932/2002). *The Philosophy of the Present*, Chicago: The Open Court Publishing Company.

Mead, G. H. (1938/1972). *The Philosophy of the Act*, edited by Morris, C. W. with Brewster, J. M., Dunham, A. M. and Miller, D., Chicago: University of Chicago Press.

Meirieu, P. (2007). *Pedagogie: Le devoir de resister (Education: The duty to resist)*, Issy-les-Moulineaux ESFediteur.

Miles, R., Lemon, N., Mitchell D. M. and Reid J-A. (2016). The recursive practice of research and teaching: Reframing teacher education, *Asia-Pacific Journal of Teacher Education*, 44(4), 401–414.

Miller, D. L. (1989). *Lewis Mumford: A Life*, New York: Weidenfield & Nicholson.

Miller, D. L. (1990). Consciousness, the attitude of the individual and perspectives, in Gunter, P. A. Y. (Ed.) *Creativity in George Herbert Mead*, Maryland: University Press of America, 3–44.

Mills, A. J., Durepos, G. and Wiebe, E. (2009). *Encyclopedia of Case Study Research*, Thousand Oaks: Sage.

Minsky, M. (1986). *The Society of Mind*, New York, London, Toronto, Sydney: Simon & Schuster Paperbacks.

Moll, L. C. (2014). *L. S. Vygotsky and Education*, London, New York: Routledge.

Morris, C. W. (Ed.). (1934/1962). *Mind, Self & Society: From the Standpoint of a Social Behaviourist*, Chicago, London: University of Chicago Press.

Mumford, L. (1967). *The Myth of the Machine: Technics and Human Development*, New York: Harcourt, Brace and World.

Nakata, M. (2007). The cultural interface. *The Australian Journal of Indigenous Education*, 36, Supplement, 7–14.

NAPLAN. (2017). National Assessment Program for Literacy and Numeracy, Canberra: ACARA accessed at www.nap.edu.au/naplan/naplan.html, 7 December.

New Scientist. (2017). Go-playing super AI transcends humanity, *New Scientist*, 21 October.

Papert, S. (1980). *Mindstorms: Children, Computers and Powerful Ideas*, Brighton: Basic Books.

Papert, S. (1995). A word for learning, in Kafki, Y, and Resnik, M. (Eds) *Constructionism in Practice: Designing Thinking and Learning in a Digital World*, New York, London: Routledge, pp. 9–24.

Papert, S. (1996). *The Connected Family: Bridging the Digital Generation Gap*, Marietta: Longstreet Press.

Paterson, A. B. (2017). Paterson, Andrew Barton (Banjo) (1864–1941), *Australian Dictionary of Biography*, accessed at http://adb.anu.edu.au/biography/paterson-andrew-barton-banjo-7972, 12 December 2017.

Peirce, G. S. (2015). How to make our ideas clear, in Buchler, J. (Ed.) *Philosophical Writings of Peirce*, New York: Dover Publications, 23–41.

Peirce , G. S. (2015a). The essentials of pragmatism, in Buchler, J. (Ed.) *Philosophical Writings of Peirce*, New York: Dover Publications, 251–268.

Peirce , G. S. (2015b). Logic as semiotic: The theory of signs, in Buchler, J. (Ed.) *Philosophical Writings of Peirce*, New York: Dover Publications, 98–119.

Peters, M. A. (2015). Socrates and Confucius: The cultural foundations and ethics of learning. *Educational Philosophy and Theory*, 47(5), 423–427.

Peterson, P. (1998). Why do educational research? Rethinking our rules and identities, our texts and contexts. *Educational Researcher*, 27(4), 4–10.

Piaget, J. (1971). *Structuralism*, London: Routledge and Kegan Paul.

Pinker, S. (1994). *The Language Instinct: How the Mind Creates Language*, New York, London, Toronto, Sydney: HarperCollins Publishers.

Pratt, N. (2016). Neoliberalism and the (internal) marketisation of primary school assessment in England. *British Educational Research Journal*, 42(5), 890–905.

Purvis, J. (2002). *Emmeline Pankhurst: A Biography*, London, New York: Routledge.

Ravitch, D. (2013). *Reign of Error: The Hoax of the Privatisation Movement and the Danger to America's Public Schools*, New York: Alfred A Knopf.

Resnik, M. (2017). *Lifelong Kindergarten: Cultivating Creativity through Projects, Passion, Peers and Play*, Cambridge MA: The MIT Press.

Rios-Aguilar, C., Kiyama, J. M., Gravitt, M. and Moll, L. C. (2011). Funds of knowledge for the poor and forms of capital for the rich? A capital approach to examining funds of knowledge, *Theory and Research in Education*, 9(2), 163–184.

Robinson, K. (2017). *Out of Our Minds: The Power of Being Creative*, Third Edition, West Sussex: John Wiley and Sons.

Rorty, R. (1979/2009). *Philosophy and the Mirror of Nature*, Thirtieth Anniversary Edition, Princeton, Oxford: Princeton University Press.

Roth, W-M. (2011). *Passibility: At the Limits of the Constructivist Metaphor*, Dordrecht: Springer.

Russell, K. (2007). Feminist dialectics and Marxist theory, *Radical Philosophy Review*, 10(1), 33–54.

Rutkowski, L. and Rutkowski, D. (2016). A call for a more measured approach to reporting and interpreting PISA results, *Educational Researcher*, 45(4), 252–257.

Samarji, A. and Hooley, N. (2015). Inquiry into the teaching and learning practice: An ontological-epistemological discourse, *Cogent Education*, 2(1). doi: 10.1080/2331186X.2015.1120261.

Scholl, R., Nichols, K. and Burgh, G. (2016). Connecting learning to the world beyond the classroom through collaborative philosophical inquiry, *Asia-Pacific Journal of Teacher Education*, 44(5), 436–454.

Searle, J. R. (1998). *Mind, Language and Society: Philosophy in the Real World*, New York: Basic Books.

Seigfried, C. H. (1996). *Pragmatism and Feminism: Reweaving the Social Fabric*, Chicago, London: University of Chicago Press.

Sellar, S. and Lingard, B. (2014). The OECD and the expansion of PISA: New global modes of governance in education. *British Educational Research Journal*, 40(6), 917–936.

Shulman, J. (1991). Revealing the mysteries of teacher-written cases: Opening the black box. *Journal of Teacher Education*, 42(4), 250–262.

Singh, M. and Major, J. (2017). Conducting Indigenous research in western knowledge spaces: Aligning theory and methodology. *The Australian Educational Researcher*, 44(1), 5–19.

Stapp, H. (2009). *Mind, Matter and Quantum Mechanics*, Third Edition, Verlag, Berlin, Heidelberg: Springer.

Stenhouse, L. (1983). Research as a basis for teaching, in L. Stenhouse (Ed.). *Authority, Education and Emancipation*, London: Heinemann.

Strhan, A. (2016). Levinas, Durkheim and the everyday ethics of education. *Educational Philosophy and Theory*, 48(4), 331–345.

Taylor, R. M. (2016). Open-mindedness: An intellectual virtue in the pursuit of knowledge and understanding. *Educational Theory*, 66(5), 599–618.

Tegmark, M. (2014). *Our Mathematical Universe: My Quest for the Ultimate Nature of Reality*, New York: Alfred A. Knopf.

Tegmark, M. (2017). *Life 3.0: Being Human in the Age of Artificial Intelligence*, New York: Alfred A. Knopf.

Turkle, S. (2017). Remembering Seymour Papert, *LRB Blog*, accessed at www.lrb.co.uk, 15 February 2018.

Turner, V. (1967). *The Forest of Symbols: Aspects of Ndembu Ritual*, New York: Cornell University Press.

UN. (2007). United Nations Declaration on the Rights of Indigenous Peoples, accessed at www.un.org/esa/socdev/unpfii/documents/DRIPS_en.pdf, 9 November, 2017.

Van Maanen, J. (1988). *Tales of the Field: On Writing Ethnography*, Chicago, London: The University of Chicago Press.

Veresov, N. (2005). Marxist and non-Marxist aspects of the cultural-historical psychology of L. S. Vygotsky. *Outlines*, 1, 31–49.

Watts, E. J. (2017). *Hypatia: The Life and Legend of an Ancient Philosopher*, New York: Oxford University Press.

Weiner, G. (1995). *Feminism in Education: An Introduction*, Buckingham: Open University Press.

Weizenbaum, J. (1976/1984). *Computer Power and Human Reason: From Judgment To Calculation*, San Francisco: W. H. Freeman.

White, J. (2018). Philosophy and teacher education in England: The long view. *British Journal of Educational Studies*, published online 24 January 2018, accessed at https://doi.org/10.1080/00071005.2018.1426830, 4 February 2018.

Wilkins, J. (2012). The spectre of neoliberalism: Pedagogy, gender and the construction of learner identities. *Critical Studies in Education*, 53(2), 197–210.

Williams, R. (1989). *Resources of Hope: Culture, Democracy, Socialism*, London, New York: Verso.

Wilson, E. O. (2017). *The Origins of Creativity*, Harmondsworth: Penguin, Random House UK.

Wilson, J. D. (Ed.). (1932/1984). *Matthew Arnold: Culture and Anarchy*, Cambridge, UK: Cambridge University Press, p. 72.

Wilson, T. S. and Ryg, M. A. (2015). Becoming autonomous: Nonideal theory and educational autonomy. *Educational Theory*, 65(2), 127–150.

Wittgenstein, L. (1922/2016). *Tractatus Logico-Philosophicus*, New York: Dover Publications.

Wittgenstein, L. (1958). *Philosophical Investigations*, Third Edition, Anscombe, G. E. M. (Translator), Upper Saddle River: Prentice Hall.

Wollstonecraft, M. (1994). *Maria, or The Wrongs of Woman*, with an Introduction by Mellor, A. K., New York, London: W. W. Norton & Company.

Wyness, M. and Lang, P. (2016). The social and emotional dimensions of schooling: A case study in challenging the 'barriers to learning'. *British Educational Research Journal*, 42(6), 1041–1055.

Yunkaporta, T. (2009). Aboriginal pedagogies at the cultural interface, PhD thesis: Queensland: James Cook University.

Yunkaporta, T. and Kirby, M. (2011). Yarning up Indigenous pedagogies: A dialogue about eight Aboriginal ways of learning, in Purdie, N., Milgate, G. and Bell, A.R. (Eds) *Two Way Teaching And Learning: Toward culturally reflective and relevant education* Melbourne: ACER Press, 205–214.

Zizek, S. (2014). *Absolute Recoil: Towards A New Foundation Of Dialectical Materialism*, London, New York: Verso.

INDEX

Locators in **bold** refer to tables and those in *italics* to figures.